The Green Cheat

Scientists and everyday people challenge the climate change narrative

TWR — Jan Webmedia

The Green Cheat

Scientists and everyday people challenge the climate change narrative.

Common sense, words of wisdom, scientific evidence

Edited by Veronica Finch

TWR

—

Jan Webmedia

For all our children and grandchildren.
May future generations be blessed with peace, freedom and truth.

— — —

A massive thank you to everyone who has contributed to this book.
It's been good to see how many were prepared to write down their concerns and
share with the world.
Special thanks to Ian Dickinson for his amazing and reliable help
with proofreading.
Thank you to my husband for his dedicated help during the progression of this
book, for his patience and persistence. What a journey it's been!

ISBN: 978-1-0685069-0-1

First published in 2024

This edition published 2024 by TWR
An imprint of Jan Webmedia

Design and formatting: Jan Webmedia

janwebmedia.uk

Contents

warming—*Dr John Gideon Hartnett*

There appears to be no credible scenario where driving emissions of CO_2 to zero by the year 2050 would avert a temperature increase of more than a few hundredths of a degree centigrade. The immense costs and sacrifices involved would lead to a reduction in warming approximately equal to the measurement uncertainty. It would be hard to find a better example of a policy of all pain and no gain.
—*Dr Richard Lindzen, Dr William Happer, Dr William A. van Wijngaarden*

Introduction

Dear reader. You are holding in your hands a book with 35 essays written by almost as many authors, varying in style, content and level—some lighthearted and entertaining, some with a deep look behind the smoke screen, some philosophical, some attempting to open the reader's eyes through fiction and others providing solid scientific evidence, sometimes requiring a high level of comprehension. The different voices make up an interesting and accessible read for people from any educational or social background.

What all the essays have in common is an understanding of the real situation concerning climate change. We are not being fooled (any more). We have seen through the fraudulent narrative. We've had enough of the nonsense.

Among us are analysing scientists, mums and dads with common sense, wise elderly people with life experience, concerned citizens seeking the truth.

The essays not only explain why the current climate change narrative is wrong, they also disclose the ugly truths about the origin of the so-called green agenda. It's scary, yet important to know that we aren't only dealing with unscientific and unproven theories that are being strongly propagandised in mainstream media and education, but also with very dark motives behind it all.

If you are new to questioning climate change, be prepared to view the world from a very different perspective. After reading *The Green Cheat* there is no going back. I trust that sooner or later we will witness victory over the corruption of environmental issues that has held the world in grip for decades, with the rise of climate insanity and hysteria in recent years. The time has come for it all to end.

This book doesn't have to be read from cover to cover. Start reading wherever you want, skip some parts, flip back and forth as you wish. Read what is accessible to you at the moment and come back later for the rest.

If you have enjoyed reading *The Green Cheat*, please spread the word. Lend the book to a family member, offer it to your local library, or gift a new copy to a friend. Small efforts like this contribute to bringing down the green tyranny and ending ecological corruption.

Veronica Finch

I have just heard on BBC Radio 2 that we only have five years to save the planet from global warming, I remember Greta Thunberg saying those exact words ten years ago! If we are now having the hottest spring on record, why am I feeling so cold? It's time people stopped listening to the BBC, you can tell just by looking out of your window what the weather is like. For information, you do not have to have a TV licence unless you watch live TV. Dump the BBC and watch your own DVDs for free. That way, you don't have to put up with their lies and one-sided propaganda.

—*John Willis*

I.

Greta the Great

By Moira M. Malcolm

> Moira M. Malcolm is Scotland's most prominent anti-lock-down campaigner, and author of *Questioning Lockdown*. She was arrested while lending books from a suitcase in Dundee City Centre in January 2021. Moira thoroughly researches topics/ policies which create new tax revenues and interfere with our inalienable human rights.

I 5-year-old Swedish school girl Greta Thunberg first played truant on Fridays in August 2018. She sat outside a government building with a sign which said 'skolstrejk för klimatet' translation 'school strike for climate', demanding politicians take action to meet the Paris Agreement which was negotiated at the 2015 United Nations Climate Change Conference in Paris. Greta encouraged children throughout the world to join her protest by not going to school one day a week.

Encouraged not punished

Attending school is compulsory in Sweden, therefore you'd have expected Swedish authorities to take action to stop the miscreant schoolgirl inciting others to not go to school but this is not what happened. Instead, there followed a media frenzy supportive of school truancy as a form of protest against the government. To the educational detriment of themselves, their peers and teachers whose classes they disrupted, school pupils the world over were encouraged to join Greta's *Fridays For Climate* protests.

There was no suggestion of Greta being offered counselling even though her parents gave testimony of her stopping eating and speaking such was her fear of the earth heating up. There was no suggestion of the distressed girl being coaxed to attend school instead of sitting on concrete outside the Swedish parliament. Instead, Greta was applauded in the media and schoolchildren in other countries were incited to become equally hysterical and join her in solidarity by missing school.

For example, in Scotland, on 15 March 2019, a BBC article titled 'Scotland's school pupils in second mass climate strike' even guaranteed beforehand, 'A number of Scottish councils said children would not face punishment if they took part in the protests with their parents' permission. They include: Argyll and Bute, Dumfries and Galloway, Dundee, Eilean Siar, Fife, Glasgow, Highland, and Renfrewshire.' The article also states, 'First Minister Nicola Sturgeon tweeted her support of the pupils "taking a stand on climate change"'.

When Greta's *Schools 4 Climate Action* protests received worldwide publicity and government officials encouraged pupils not to attend school on Fridays in order to protest against the government, it became clear this was most definitely a government-backed agenda. Schoolchildren were being used as government pawns to promote the UN climate change agenda and no one was being used more than Greta Thunberg herself.

INTERESTING RELATION

One of the first scientists to make hysterical predictions about global warming, claiming CO_2 emissions cause the earth's surface temperature to increase, was none other than

Swede Svante Arrhenius, a distant relative of Greta! What are the chances?

It is Greta's grandfather, Olof Thunberg who is related to Svante Arrhenius, who we learn from Eugenics Archives[1] was a board member for the Swedish Society for Racial Hygiene. The definition of 'racial hygiene' in the Cambridge dictionary is: 'the idea that a race of people can be kept pure by not allowing people who are considered inferior (=of less value) to have children.'

Greta's immediate family are all involved in the arts. Her grandfather, Olof Thunberg was a famous actor and director in Sweden. Greta's mother and father are both trained actors. There's certainly a lot of knowledge and experience within the family to act out a nicely crafted script around climate change hysteria. It's likely the whole thing was scripted from the outset, that Greta was sent to sit outside the government building with a placard and told to play her part.

INVITES GALORE

Whenever billionaires attend conferences, e.g. World Economic Forum (WEF) in Davos or fanfare United Nation climate change conferences, you can guarantee Greta Thunberg will be there. Greta is given a platform to rant angrily about not enough being done to prevent climate change. One can imagine those present rubbing their hands underneath the tables because she is saying exactly what they want her to say.

On the back of relentless propaganda (and silencing scientists who disagree), a new financial market, trading carbon emissions, is in full swing. This enormous new market has the effect of duping an entire generation into championing their own serfdom.

---

1 Profile of Svante Arrhenius, Eugenics Archives [website], <https://www.eugenicsarchive.ca/connections?id=531f7ac0132156674b000204>, accessed 2 September 2024.

Jollies for Greta

There aren't many teenagers who refuse to travel by plane due to their belief regarding CO_2 emissions, who are then transported by racing yacht across the Atlantic Ocean, but this is how precious Greta is to the billionaire class who pull her strings. Greta's 'activism' is relentless. At Davos in January 2019 she said this: 'I don't want you to be hopeful. I want you to panic. I want you to feel the fear I feel every day. And then I want you to act.'

Greta wants you to act by consenting, even demanding, to be enslaved in an open prison 15-minute city, to act to ensure every purchase you make is tracked and traced (carbon credits awarded or subtracted), for every flight (except private jets) to be grounded, for meat to be a luxury only for the Davos set.

Don't fall for it. Greta Thunberg's distant relative, Svante Arrhenius was a puppet of the state and so is she.

The establishment brazenly asserts that man-made CO_2 is the main 'control knob' of global climate, an absurd assertion which is unsupported by empirical data and flies in the face of common sense.

—*Douglas Brodie*

II.

Debunking the climate change hoax, part one

By Douglas Brodie

INTRODUCTION

I am convinced that 'climate change' is a hoax from many years of campaigning against it and from the establishment's globally-coordinated excesses of recent years which have confirmed it to be wantonly abusive, mendacious and conspiratorial in its dealings with the people.

This paper aims to show that anyone who still believes in dangerous man-made CO_2 global warming (aka 'climate change') urgently needs to reassess their views before the dictatorial Net Zero oppressions and privations being inflicted on us under the pretence of 'tackling climate change' become unacceptably onerous and destructive.

It exposes the blatant falsehoods of the establishment's climate change narrative and you don't have to be a science egghead to see through their 'really very stupid' deceptions. You only need to open your mind to the sad reality that almost everything the duplicitous establishment and paid-for mainstream media have told us about 'climate change' is a lie. The simple explanations and facts in this paper will then allow you to see through all of their outrageously false 'climate change' propaganda and brainwashing.

I have given up trying to reason with closed-minded, electorate-betraying politicians. I am now reaching out to ordinary people, many perhaps lacking the capability to challenge what is going on but becoming more and more suspicious that they are being horribly deceived and abused for ulterior political motives on 'climate change' and other globalist machinations.

We mustn't allow the establishment's junk science, anti-humanity climate policies to lead us ever deeper into energy infrastructure ruination.

THE ORIGINS OF THE 'CLIMATE CHANGE' HOAX

This exposes the seldom-mentioned anti-capitalism, anti-democracy and even anti-humanity origins of the scientifically corrupt, Malthusian, horribly politicised, horribly entrenched and ruthlessly enforced climate change hoax.

Concerns about the impact of humanity on the environment and planetary resources were raised long ago by Thomas Malthus (1766–1834). In recent times, various bad actors have weaponised these concerns for dark ulterior motives. One such was the late billionaire socialist Maurice Strong who set up the UN's Intergovernmental Panel on Climate Change (IPCC) in 1988 and its Earth Summit in 1992, origin of the Agenda 21 template for authoritarian UN world governance and global wealth redistribution. He is remembered for his infamous saying:

'Isn't the only hope for the planet that the industrialized civilizations collapse? Isn't it our responsibility to bring this about?'*

The Machiavellian Strong set about creating a false problem

---

* This quote is part of a book Strong planned to write, in which he was able to say the 'quiet part out loud', to express his real views through fiction in a way he couldn't in a public speech. *(ID)* based on false science, drawing on the 1979 Charney Report (since debunked) which set the ball rolling on the conveniently abstruse hypothesis of man-made CO_2 global warming. He set up a system of democracy-bypassing bureaucrats to get the developed countries to deindustrialise and make them pay, inspired by this Club of Rome statement:

'The common enemy of humanity is man. In searching for a new enemy to unite us, we came up with the idea that pollution, the threat of global warming, water shortages, famine and the like would fit the bill. ... The real enemy then is humanity itself.'

The UN IPCC's climate change skulduggery started in earnest in their 1995 report when lead author Ben Santer mendaciously claimed that global temperature data showed a 'discernible human influence'. Santer's claimed human influence has to this day never been reproduced and the establishment's never-proven man-made CO_2 global warming hypothesis has been disproved by many studies, such as the Siegel climate change challenge[1]. All we ever get from the UN is bluster, not proper science, with the UN Secretary General ranting about 'global boiling' and the UN climate chief making an unscientific fool of himself by emoting that we only have two years to save the world.

Big Industry and Big Money enthusiastically joined the bandwagon to grab a share of the heavily-subsidised (at hard-pressed taxpayer expense) $multi-trillion climate change bonanza. The party was also joined by the unaccountable Davos World Economic Forum and multi-billionaires with unlimited budgets for propaganda, censorship and brainwashing such as Bill Gates, George Soros and the Rockefellers who are

---

1 David Siegel, 'Siegel climate change challenge', 28 April 2024, *YouTube* [website], <https://www.youtube.com/watch?v=QCO7x6W6iwc>, accessed 30 August 2024.

using the climate change hoax to manipulate public policy towards their own self-centred goals.

THE PROOF THAT 'CLIMATE CHANGE' IS A HOAX

The establishment brazenly asserts that man-made CO_2 is the main 'control knob' of global climate, an absurd assertion which is unsupported by empirical data and flies in the face of common sense. Temperatures were congenially warmer than now, which goes towards explaining why Orkney flourished so amazingly in these early times.

A subsequent period of cooling was followed by the Minoan Warm Period then the Roman Warm Period, when grapes were cultivated along Vine Lane in Newcastle (Hadrian's Wall). Was this warm period caused by a spike in atmospheric CO_2 levels due to heavy Roman chariot activity?! No, this reconstruction shows that CO_2 levels then were much lower than now. Whatever climatic influence led to these benign conditions, it gave way to the severe cold of the Dark Ages (with negligible change in CO_2 levels) which led to massive European population migrations seeking better living conditions.

The Dark Ages gave way to the Medieval Warm Period when magnificent cathedrals were built all over Europe and Vikings settled in Greenland. These benign conditions in turn gave way to the Little Ice Age (~1300 to ~1800). It got so cold that ice fairs were held on the frozen River Thames and the Vikings were forced out of icebound Greenland. Modern science then started to advance and revealed that these cold periods coincided with periods of extremely low solar activity, e.g. the Maunder Minimum.

Solar activity started to increase some time before the mid-18th century start of the Industrial Revolution and the Little Ice Age gave way to gradual but fitful global warming. This led to the establishment's late-20th century play-acting alarm about global warming, which they pretended had been caused by man-made CO_2 when it was clearly due to the strong solar activity of the Modern Maximum (see the Maunder graphic above), the strongest in centuries. The establishment shamelessly ignores this inconvenient science and brazenly spins the yarn that the global warming since the start of the Industrial Revolution has been mainly due to rising CO_2 levels.

They never mention the reality that we are living in a CO_2 famine relative to the earth's much greener, CO_2-richer yet never 'boiling' past. They even double down by fearmongering about needing to 'sequester' CO_2 from the atmosphere.

The Climategate emails leaked in 2009 from the Climate Research Unit of the University of East Anglia, a key hub of UN IPCC climate science, showed how its climate scientists behaved as bought-and-paid-for establishment puppets, e.g. when one of them gave the game away by writing:

'We have to get rid of the Medieval Warm Period.'*

This led to the creation of the fake hockey stick graph of global temperatures which falsely depicted almost no change in 1,000 years then a sudden spike in the late 1900s. It was gleefully promoted by the duplicitous UN IPCC as it portrayed the late-20th century global warming as unprecedented and gave support to their CO_2 'control knob' pretence. It took years of dogged investigation by a small band of climate realists, hindered all the way by IPCC scientists who refused to make public their data, to prove that the hockey stick graph was bogus, crafted from flawed data and statistical chi-

---

* The actual quote from the email is: 'I get the sense that I'm not the only one who would like to deal a mortal blow to the misuse of supposed warm period terms and myths in the literature' as explained here: https://wattsupwiththat.com/2013/12/08/the-truth-about-we-have-to-get-rid-of-the-medieval-warm-period. From the web page: 'As to this being a fabrication (as commentator Robert claims), no, it's a summation or a paraphrase of a long quote, something that happens a lot in history.' (ID)

canery. The book *The Hockey Stick Illusion* by Andrew Montford details the deception.

The establishment whitewashing of the Climategate scandal is now being re-enacted in the similarly biased Hallett Inquiry into the scientific malfeasances of the Covid scandal, parallel examples of establishment lying by omission to take us all for fools and paint dissenters as 'conspiracy theorists'. Their coordinated moves to cancel, censor and even jail dissidents prove that they themselves are the conspirators.

The pretence that rising levels of atmospheric CO_2 lead to dangerous global warming is debunked by studies and reconstructions of what happened in the recent and long-ago past, as described in 'Net Zero Shock: Carbon Dioxide Rises AFTER Temperature Increases, Scientists Find' (dailysceptic.org). These show that the increase in atmospheric CO_2 follows the rise in global temperature rather than coming before it and causing it, i.e. the exact inverse of the establishment's CO_2 climate 'control knob' pretence.

See also 'Smoke, mirrors and CO2 emissions' (conservativewoman.co.uk) describing two hoax-busting spells of global cooling since the mid-1940s which happened despite ever-rising atmospheric CO_2, the first in the 1960–70s so worrisome at the time as to cause a new ice age scare, plus a catalogue of 'smoke and mirrors' obfuscations and deceptions in the establishment's mendacious climate change narrative.

Further evidence that CO_2 global warming is a hoax is given by the abject failure of the establishment's computer climate models to give credible global temperature predictions. Here they are failing miserably in graphs of predictions versus observations in 2015 and even worse in 2022. Evidence is also given by the establishment's Chicken Licken predictions of climate-related catastrophes which have failed to come to pass, now going back 20 years and more. Likewise, the establishment's make-believe 'climate emergency' has been rebutted by the World Climate Declaration signed by thousands of independent scientists.

The establishment's climate modelling failures are easily explained by the many studies[2], which show that the warming effect of CO_2 is already 'saturated' and that even a far-off doubling of the concentration of CO_2 in the atmosphere (currently 420 ppm) will have minimal impact on global temperatures. The latter links to studies which show that nitrogen and methane likewise have minimal impact on climate. The establishment ignores all such inconvenient science and instead uses their own junk science as pretext to wage war on farming in pursuit of the dystopian goals of UN Agenda 2030 which no electorate has ever voted for.

Few people realise that the UN IPCC is only mandated to study the risks of human, not natural influences on climate. They never admit this objectivity-wrecking restriction to the general public. It allows them to pretend that rising atmospheric CO_2 levels must be the problem because they disregard almost everything else. They ignore solar variations, planetary orbital and gravitational variations, solar/ ocean-driven ENSO, PDO and AMO cycles and much more to pretend that greenhouse gases—emphasising man-made CO_2 and not even mentioning water vapour, the most important greenhouse gas—and other 'anthropogenic forcings' are the main drivers of climate.

The UN IPCC claims that atmospheric CO_2 must be the main driver of climate because the change in Total Solar Irradiance (TSI) over the course of the 11-year solar cycle is very

---

2 Example: Ariane Loening, 'More Carbon Dioxide Cannot Absorb More Infrared Radiation', 5 February 2024, *Bruges Group Blog* [website], <https://brugesgroup.com/blog/more-carbon-dioxide-cannot-absorb-more-infrared-radiation>, accessed 30 August 2024.

small. This is lying by omission because within the small TSI variation there are large variations in parts of the spectrum such as ultraviolet which affects the ozone layer, which in turn affects global temperatures.

Variations in the sun's magnetic field strength are clearly a very important natural 'control knob' of global climate.

The solar wind magnetic field affects the Earth's climate by shielding us from incoming cosmic rays when it is strong, resulting in reduced cloud cover and higher global temperatures and vice versa when it is weak. This effect has been studied by independent scientists Svensmark and Shaviv and has strong observational backing. It is explained in the recent hoax-debunking documentary *Climate: The Movie* (Martin Richard Durkin). The establishment simply ignores this inconvenient science.

Independent astrophysicist Dr Willie Soon has debunked the UN IPCC's obtuse stance on CO_2 by analysing the rural temperature record (avoiding Urban Heat Island warming bias) for the past 150 years and the corresponding changes in solar activity to reach the obvious conclusion that it is the sun, not CO_2 that drives global temperatures. Former IPCC supporter Professor Fritz Vahrenholt reached the same conclusion a decade ago as documented in his book *The Neglected Sun*.

Another establishment climate change skulduggery is the never-justified retrospective adjustment of official temperature records, always in a direction to make global warming look worse. The article 'Man Made Global Warming' (notalotofpeopleknowthat.wordpress.com) describes the adjustments made by the corrupt UK Met Office to their Had-CRUT series, to the extent that 'most of the warming since 2001 is the result of adjustments to the data'. The article 'Alterations To The US Temperature Record' (realclimatescience.com) describes the shocking adjustments to US temperature records by increasing recent temperatures and reducing past temperatures. This cheating is exposed by the fact that the official US all-time high temperature records still stand in the dust bowl years of the 1930s when, as even the UN IPCC has conceded, CO_2 levels were too low to have caused such warming.

The latest establishment climate change skulduggery concerns an event which they have hidden from the public by globally-coordinated censorship, namely the 2022 Hunga Tonga undersea volcanic eruption which injected massive quantities of water vapour, the most important greenhouse gas, high into the stratosphere. Global temperatures are now showing an unprecedented ~1 °C spike unlike any past El Nino (in modern measurements) which the establishment shamelessly claims is due to man-made CO_2. This is clearly a lie, not least because the UN IPCC's own (pseudo) science predicts CO_2 global warming at a rate of at most 0.3 °C per decade. Paul Homewood easily debunks the baseless claim that 2023 was the hottest in 125,000 years[3].

The establishment censorship of Hunga Tonga gives the game away. They don't want to admit that water vapour is a much more potent greenhouse gas than CO_2 and are reduced to spouting 'global boiling' nonsense now that global temperatures have gone from benign flatlining from 1998 to 2023, with multiple ENSO transients along the way and looking very like the waning warm phase of the verboten-to-mention Atlantic Multidecadal Oscillation (AMO), to all of a sudden breaching the UN's precious 1.5 °C limit!

---

3   Paul Homewood, 'Hottest In 125,000 Years?', 23 December 2023, *Not a Lot of People Know That* [website],
    <https://notalotofpeopleknowthat.wordpress.com/2023/12/23/hottest-in-125000-years>, accessed 30 August 2024.

Their bourgeois lifestyle was an illusion of elitism and only a temporary bribe to ensure their compliant efforts in creating and facilitating this evil scheme.

There is hope that the people who threw us under the bus for a bag full of gold may well be the ones who save us from the New World Order.

—Hannah

III.

The great green money grab

By Hannah

Hannah is the owner of an online vintage and second hand shop and former owner of a bricks and mortar vintage shop. She is all for living a sustainable life: preserving finite materials, reducing waste, eliminating toxic chemicals, etc. However, she has never believed the fake crisis currently named 'climate change' and has never profited from this unholy bandwagon that so many people have fallen over themselves to monetise.

Decades ago, I had my heart broken by my boyfriend of ten years. His parting words on the telephone included the following statements:

1. I've just got a promotion at work.
2. I'm mixing with women with careers, company cars and pensions schemes.
3. I want to go to parties on the Thames and rub shoulders with the rich and famous and you are never going to help me get there.
4. You are going nowhere in life.
5. Nothing good will ever come of you.
6. No man will ever want to marry you.
7. I only stayed with you for...
8. It's over.
9. Never contact me again.

He'd split up with me several times before, always in cruel ways, but this time I knew it was final.

This man had always shown signs of wanting the high life: he'd regularly sworn he would one day be a millionaire, he prioritised foreign holidays above home life and he had to have designer clothing and Toni & Guy haircuts. He also enjoyed thrill-seeking adventures and there was an element of one-upmanship amongst his equally successful travelling mates. I was more into vintage things, restoring furniture, DIY haircuts and prettying up the home. In fact, I once half-restored his first flat and even took two weeks off work to decorate a bedroom for him so he could rent it out to earn more money.

Why does this relate to the Green agenda? Because he now owns a very successful business, employing dozens of people, advising other businesses on how to achieve Net Zero. He also owns a business involved in solar fields. I emailed him several years ago specifically to relay what I knew about the Net Zero and climate change hoax. I pointed out that it was a trap, designed by the globalist elite, to usher in a New World Order, reduce the global population and enslave the survivors into a high-tech dystopian smart cities prison planet.

His shameless response was, 'I know that, but it's making me a lot of money.'

One of the hardest things in life is accepting that people you believed cared about you didn't care about you at all. Not just people but also governments, healthcare providers, celebrities, journalists, elites, etc. Waking up to the fact that some people or entities do not care about our welfare can be an extremely painful and disorientating phase to have to go through. Many refuse to accept it, despite overwhelming evidence, and stay stuck in a state of denial and cognitive dissonance: acceptance being too overwhelming to face.

The sad fact is that money corrupts and the Net Zero industry is creating a lot of millionaires. Once established in these 'Green' business circles, the invitations to conferences and swanky after-parties will roll in. Mixing with equally corrupt money-grabbing individuals from exploitative businesses not only encourages more of the same but also fails to offer any grounding opportunities. Thick as thieves, they swap ideas, trade secrets and toast their ill-gotten gains, away from the prying and annoyingly judging eyes of the lesser mortals who weren't lucky enough to get on the bandwagon. In the extreme, attendees will be invited to private parties at these conferences, such as Davos, and be filmed in compromising situations having had their drinks spiked. This explains the whole Epstein entrapment enterprise.

Their consciences are either severely deficient, due to hippocampus and amygdala related psychopathy, or they are acting psychopathically and managing to cope with the cognitive dissonance that must come from knowing it's all a lie but milking it anyway. My belief is there will be a lot of nervous breakdowns in Net Zero circles: good people will cave in when things really start to happen. People will speak out, they will admit their hand in this debacle and there will be many apologetic revelations from whistle-blowers. The ones who don't speak out will lose in the end: their big houses and foreign holidays will be deemed unsustainable. Trading carbon credits will only keep them in the comfort they have become accustomed to for the time being. Eventually, their status will be the same as ours: proletariat or 'useless eaters'. Their bourgeois lifestyle was an illusion of elitism and only a temporary bribe to ensure their compliant efforts in creating and facilitating this evil scheme.

So, there is some hope in that: the people who threw us under the bus for a bag full of gold may well be the ones who save us from the New World Order.

'But those who desire to be rich fall into temptation, into a snare, into many senseless and harmful desires that plunge people into ruin and destruction.' (1 Timothy 6:9)

And did my ex marry a woman better than me? No, he never even married. Instead, he has strung another woman along for years, twenty-five years in fact. I confess that I looked up their home once and found its estate agent's listing when they were selling up. It was a strange experience to see the home he created without me. I say home but it didn't look much like a home: more like a second home where people stayed occasionally, slightly more appealing than student digs. The walls were all whitewashed, the kitchen appeared unused and there were barely any decorations, other than buddhas on the mantlepieces. No character, no soul.

While he is prioritising making money and going on holidays four times a year, I am focusing on creating a lovely and characterful home, and becoming as self-sufficient as possible. My main priority, however, over the past almost twenty years has been to expose the various agendas and tactics being used against us. While he has no doubt accrued lots of friends and a wide and worldly social circle, I have been vilified and abused. But I would rather stand alone than stand amongst a corrupt crowd. And, in time, when his tan has faded due to foreign travel being made possible for only the elites, and when he is forced to confine himself to his barren home during climate lockdowns, and when he can only buy food with his RFID wrist chip and when he has time on his hands to ponder the role he played in everyone's confinement, I will be enjoying my home and all of my hobbies, I will be eating my own vegetables from my garden and I will have no regrets, being on the right side of history.

Why are certain people obsessed with the idea that there are 'too many people' on this beautiful planet? In 2022 the United Nations stated that: 'More than a third of 50 recently surveyed Nobel laureates cited "population rise/ environmental degradations" as the biggest threat to humankind.' The equally evil World Economic Forum stated in 2018: 'Even as birth rates decline, overpopulation remains a global challenge.'*

Has the alarming term 'overpopulation' been used to condition us into believing that humanity itself is a problem?

We are often told that overpopulation is the reason for climate change. It is difficult to find other voices in mainstream media. Scientists who come to a different conclusion are harassed, cut off from grants and have no access to mainstream media.

If the Earth was overpopulated, wouldn't there be a problem with space on this planet?

—*Joseph Lang*

*   *United Nations* [website], <https://un.org/en/desa/population-growth-environmental-degradation-and-climate-change> and *World Economic Forum* [website], <https://weforum.org/agenda/2018/04/almost-everywhere-people-are-having-fewer-children-so-do-we-still-need-to-worry-about-overpopulation>, accessed 19 September 2024.

IV.

The green route to neofeudalism

By Niall McCrae

As laws multiply, liberties diminish. In the twentieth century, dictators pretended that their regime was for citizens' safety, but now it's to 'save the planet'. Green virtue may be emphasised, but there is nothing natural about comprehensive digital surveillance, synthetic food, mRNA vaccines and Smart Cities. This is the Great Reset, as promoted by the World Economic Forum, King Charles and other prominent figures.

Since the authoritarian lurch in the Covid-19 'pandemic', sceptics of conservative or right-wing bent have warned of a deceitful imposition of communism. Others perceive the rise of fascism. As socialist Simon Elmer explained, on observing the massive transfer of wealth from ordinary people to global corporations (particularly Big Tech, Big Pharma, and the investment bankers Vanguard and BlackRock), fascism is not defined by 'blood and soil' nationalism or legions in jackboots, but by Benito Mussolini's fusion of government and industry to create a totally ordered society.

Fascism and communism are simplistically regarded as polar opposites, but they overlap as schema of collectivisation and control. Both harness the people in a way that potential dissent or revolt are minimised, while adversities such as economic hardship are presented as a struggle against an outer or inner enemy. There is much in George Orwell's writing on the ordering and compartmentalising of ordinary people: some more equal than others, as the privileged pigs lived in *Animal Farm*.

In 1957 Ayn Rand's novel *Atlas Shrugged* appeared. A Russian emigrant to the USA, Rand was writing during the post-war consumerism of the American Dream, a stark contrast with the Soviet system of her youth. The book was highly influential for its message of individual liberty prevailing over stifling collectivism. But Rand's idealism was going from one extreme to the other. The only loyalty was to self, not unreasonably characterised by the phrase 'greed is good'. Unbridled free-market capitalism is as destructive as communism in destroying traditional society and its allegiances to faith, flag and family.

Nonetheless, capitalism needed its defenders, at a time when Marxist ideology was marching through the educational institutions of the West. Milton Friedman, the Nobel Prize-winning economist who inspired Ronald Reagan and Margaret Thatcher, explained how collectivist polity denies people freedom and choice. In his book *Freedom to Choose* (1980), co-authored with his wife Rose, Friedman described the paternalism of FDR Roosevelt's New Deal in the 1930s:

'The members of FDR's brain trust were drawn mainly from the universities—in particular, Columbia University. They reflected the change that had occurred earlier in the intellectual atmosphere on the campuses—from belief in individual responsibility, laissez-faire, and a decentralized and limited government to belief in social responsibility and a centralized and powerful government. It was the function of government, they believed, to protect individuals from the vicissitudes of fortune and to control the operation of the economy in the general interest, even if that involved government ownership and operation of the means of production.'

Although Friedman was against an over-governed society, he was an early advocate of a basic universal income. In 1962 be called for a negative income tax for the unproductive minority, developing this idea in *Freedom to Choose*. Thatcher and Reagan strove to reduce the role of government, but they were criticised for policies that caused recession and unemployment. The medicine tasted foul, but it seemed to work: capitalist economies boomed in the 1980s. Friedman's idea of reverse taxation was too radical, but it was kept alive in academic debate.

Illustrating the folly of the Left versus Right paradigm in politics, universal basic income was also favoured by Marxists. In 1966 two Columbia sociologists sought to create an egalitarian society by overwhelming federal assistance programmes, which they perceived as parsimonious. Richard Cloward and Frances Fox Pliven, inspired by militant Chicago professor Saul Alinsky, devised a revolutionary problem-reaction-solution strategy. Poverty and debt would be deliberately increased, wreaking social and economic chaos. States would then be relieved of their welfare burden, which would pass to federal administration, resulting in a basic universal income. The Cloword-Pliven strategy was influential in the Democrat Party, where it fitted the nurturing of dependent client groups.

Like Friedman, Cloward and Pliven were writing at a time when most people worked, albeit more in the peaks and fewer in the troughs of economic fortunes. Friedman's reverse taxation was intended for a prevailing minority of unproductive citizens, but unwittingly it would lead to total dependence on the state. This is the reality now faced by a global multitude, as the WEF and political parties push universal basic income, to be facilitated by a central digital currency.

In the near future most of the current workforce will be made redundant by automation and artificial intelligence.

Yuval Noah Harari, advisor to the WEF, described a mass of 'useless eaters', who spend their time playing video games. This brings to mind the Morlocks of HG Wells' dystopian *Time Machine*, an underclass that lives underground and occasionally breaks into the luxurious world of the Eloi to plunder. In the emerging two-tier society of the twenty-first century the masterclass will use advanced technology to keep the minions in check.

Having got so far in this chapter, readers may be wondering how this relates to the climate crisis and green dogma. The emerging new world order is a technocracy, which entails total control of population and resources. The quality and quantity of humankind will be manipulated by the global predators, who want to save the planet for themselves. The claimed climate crisis is nothing but an excuse to achieve this.

Dramatic depopulation is envisaged, continuing a project that has already reduced the birth rate in white Western society far below replacement level. This misanthropic mission can be traced back to the nineteenth century, not to the Industrial Revolution that is blamed for the beginning of manmade global warming, but to the most important book in modern history: Charles Darwin's *Origin of the Species*.

The theory of evolution invalidated holy scripture as historical truth, and removed the special status of human beings (atheist–materialist ideology and regimes would thus treat people as cattle). Survival of the fittest was a tenet of evolutionary nature, but fearing the relentless growth of the uncouth and uneducated masses below them, the upper class feared that society was propagating the unfit.

Leading scholar Francis Galton, Darwin's cousin, established eugenics—'the science of improving the human stock'. The Eugenics Society drew membership throughout the intelligentsia. In 1910 playwright George Bernard Shaw considered how to dispose of large numbers of undesirable people efficiently: 'I appeal to the chemists to discover a human gas that will kill instantly and painlessly'. The First World War was effectively a cull of working class men, who were gaining strength through the labour movement. Eugenics advanced in the 1920s, when California began forced sterilisation of the mentally subnormal, culminating in the Third Reich's murderous racial hygiene. After such notoriety, eugenics went backstage, but the plot continued.

In 1946, as the first director of the United Nations Educational, Scientific & Cultural Organisation, eugenicist Julian Huxley, wrote on the agency's purpose:

'At the moment, it is probable that the indirect effect of civilisation is dysgenic, instead of eugenic, and in any case it seems likely that the dead weight of genetic stupidity, physical weakness, mental instability, and disease-proneness, which already exist in the human species, will prove too great a burden for real progress to be achieved. Thus even though it is quite true that any radical eugenic policy will be for many years politically and psychologically impossible, it will be important for UNESCO to see that the eugenic problem is examined with the greatest care, and that the public mind is informed of the issue at stake so that much that now is unthinkable may at least become thinkable.'

In 1952 John D Rockefeller III, whose family's foundation had bankrolled the eugenics movement from the outset, established the Population Council (sharing offices with the American Eugenics Society, which it absorbed in 1972). Frederick Osborne, Population Council president, explained that 'eugenic goals are most likely to be attained under a name other than eugenics'.

Although population was stabilising in the West, globally it was rising steeply. Chairing the Eugenics Society from 1959 to 1962, Huxley saw the potential of ecological concern for a rebranding of eugenics. And so the misanthropic mission was cloaked in green. In 1961 Huxley co-founded the World Wild-life Fund with Prince Bernhard of the Netherlands, Prince Philip of Great Britain and Godfrey A Rockefeller. Nature conservation was the foot in the door for the catastrophism that would eventually justify the United Nations' Agenda 21 and Net Zero.

1968 is a year associated with emancipatory activism, but in retrospect it was more significant for the birth of the climate cult. The two major developments in this Anno Domini, not widely reported in mainstream media, were publication of *The Population Bomb* by Paul Ehrlich and the founding of the Club of Rome. Ehrlich's book warned of imminent famine, war and disease due to overpopulation, predicting that four billion would perish by the 1980s. The Club of Rome, established at the Rockefellers' estate in Italy, used dire prophecy to exert influence over the UN, casting human beings as a cancerous tumour on Earth.

You may have seen posters proclaiming climate change as a 'feminist issue'. In 1916 the birth control clinic Planned Parenthood was founded in New York by Margaret Sanger. In her book *Pivot of Civilization* Sanger warned of 'weeds over-running the human garden'; she urged segregation of 'morons, misfits and the maladjusted' and 'elimination of inferior races'. Sanger's message was as bad as Adolf Hitler's *Mein Kampf*, yet Planned Parenthood has continually received support from American politicians. Since the feminist wave of the 1960s the slogan 'my body, my choice' has been a rallying cry. Abortion was thus cemented to the core of feminism, while feeding a huge termination industry.

Computer salesman Bill Gates, whose father William H Gates was head of Planned Parenthood, shares his wealth through supposedly philanthropic funding of global health projects. Initially focusing on reproductive health (i.e. contraception and abortion), he moved into vaccination. In a TED talk in 2010, Gates let the cat out of the bag:

'The world today has 6.8 billion people. That's headed up to about nine billion. If we do a really good job on new vaccines, health care, reproductive health services, we could lower that by perhaps 10 or 15 percent'.

Having captured universities, regulators and media by predatory funding, the Bill & Melinda Gates Foundation was at the forefront of the Covid-19 scam. As a key player in the World Health Organisation, Gates promotes universal vaccination by novel mRNA technology, which has potential use for digital tracking and hormonal control. Former British prime minister Boris Johnson's father Stanley, who worked for the World Bank, suggested in a Guardian interview in 2012 an ideal UK population of 10–15 million.

Western countries are nudging their citizens into accepting euthanasia, with the persuasive euphemism of 'assisted dying'. Dementia, for example, is portrayed in the media as a fate worse than death. Some jurisdictions now permit 'assisted dying' for people with mental conditions, despite uncertain consent. US president Joe Biden's bioethicist Ezekiel Emanuel proposes a 'death panel' to determine the value of life, denying treatment to anyone deemed unworthy. Emanuel wants to overturn the Hippocratic Oath.

The Great Reset promoted by the technocratic World Economic Forum, UN Agenda 21 and the tightening ratchet of Net Zero policies are profoundly eugenicist. The motive, of course, is control. There is nothing green about this unscientific and profoundly unethical assault on humanity. Resistance is imperative, and that depends on awakening more people to their plight, before it is too late.

We should think of new structures for society, such as decentralisation, local communities, self-sufficient food growing, off-grid living, gathering edible wild plants, cooking with stoves and fireplaces instead of depending on electricity, having fields for growing berries and fruit trees, keeping chickens for eggs, cows for milk, collecting rain water, growing sprouts and mushrooms at home, holistic treatment and vitamin D3, learning and teaching practical skills such as knitting and furniture making, as well as reading the Bible, fearing God and following Christ.
—*Sara Kristensen, Denmark*

V.

The intellectual's new car—a fairy story

By Amaranth Briarball

Amaranth Briarball (pseudonym) is a former official of an international organisation who is thankful to have left that body before its policies became intolerable, clearly aimed at enslaving rather than assisting humanity.

Once upon a time there was an intellectual so exceedingly fond of the news that he spent all his time reading newspapers or listening to the BBC. He never observed the outside world, and he never spoke except to show off the great knowledge he had acquired. He had a hypothesis for every hour of the day, and instead of saying, as one might, about any other intellectual, 'Mr Intellect is in a meeting', here they always said, 'Mr Intellect is in his reading room.'

In the great city where he lived, life always seemed tolerable apart from the excess CO_2 the news kept telling him he was suffering from. Every day many strangers flew into the city's airport, and among them one day came two swindlers. They let it be known they were car manufacturers, and they said they could make the most magnificent car imaginable. Not only was it the cleanest car you could wish for and would run indefinitely without ever needing fuel or emitting CO_2, but it had a wonderful way of stopping dead, or even bursting into flames, if anyone unfit for his office, or unusually stupid, happened to ride in it for more than a few kilometres or failed to see how exceptionally clean it was.

'This would be just the car for me,' thought the intellectual as he read about the car dealers in the news. 'If I owned this car and lent it to others, I would be able to discover which men in my city are unfit for their posts. And I could tell the wise men from the fools. Yes, I certainly must order this car right away.' He paid the two swindlers a large sum of money to start work at once.

They set up a manufacturing plant and proceeded to build several cars. The intellectual ventured for once out of his reading room and observed what looked like power lines entering the plant, smoke rising from the flues in the roof, and bags of lithium and money marked 'government property' being carried in through a side door. But he felt slightly uncomfortable when he remembered that those who were unfit for their position would not be able to see how magnificent and how clean, in all respects, his car was. It couldn't have been that he doubted himself, yet he thought he'd rather send someone else to see how things were going. By now the whole town knew about the car's miraculous properties, and all were impatient to find out how stupid their neighbours were.

'I'll send my honest old colleague to the plant,' the intellectual decided. 'He'll be the best one to tell me what the car is like, for he's a sensible man and no one does his duty better.'

So the honest old colleague went to the plant where the two swindlers sat working away at their clean, fuel-free vehicles.

'Heaven help me,' he thought as his eyes flew wide open, 'This plant is using lots of electricity and bags of lithium marked "Mined by children in central Africa".' But he did not say so.

Both the swindlers begged him to be so kind as to come near to approve the excellent design and the purity of the manufacturing. The poor old colleague stared as hard as he dared. He could hardly see anything but power lines, smoke, bags of lithium, and a large chest bursting with money. 'Heaven have mercy,' he thought. 'Can it be that I'm a fool? I'd have never guessed it, and not a soul must know. Am I unfit to be an intellectual? It would never do to let on that I can see the pollution and the fraud.'

'Don't hesitate to tell us what you think of it,' said one of the car manufacturers.

'Oh, it's beautiful—it's enchanting.' The old colleague peered through his spectacles. 'Such a sleek design, what a finish! I'll be sure to tell my esteemed colleague how delighted I am with it.'

'We're pleased to hear that,' the swindlers said. They proceeded to name all the paint colours and to explain the high-tech design. The old colleague paid the closest attention, so that he could tell it all to the intellectual. And so he did.

The swindlers at once asked for more money, more lithium and more power, to get on with the manufacturing. But nearly all the money went into their pockets, though they worked at their manufacturing as hard as ever.

The intellectual presently sent another trustworthy colleague to see how the work progressed and how soon his car would be ready. The same thing happened to him that had happened to the first. He looked and he looked, but there was little to see except pollution and money bags.

'Isn't it a beautiful piece of equipment?' the swindlers asked him, as they displayed and described their clean car.

'I know I'm not stupid,' the man thought, 'so it must be that I'm unworthy of my good office. That's strange. I mustn't let anyone find it out, though.' So he praised the cleanliness he did not see. He declared he was delighted with the beautiful paint colours and the exquisite design. To the intellectual he said, 'It held me spellbound.'

All the town was talking of this splendid car, and the intellectual wanted to see it for himself while it was still in the manufacturing plant. Attended by a band of chosen men, amongst whom were his two old trusted colleagues who had been to the plant, he set out to see the two swindlers. He found them manufacturing with might and main, enveloped in power lines, smoke and lithium powder.

'Magnificent,' said the two colleagues already duped. 'Just look, old friend, what colours! What a design!' They pointed to the clean car, each supposing that the others could see the cleanliness of it all.

'What's this?' thought the intellectual. 'I can't see anything clean. This is terrible! Am I a fool? Am I unfit to be an intellectual? What a thing to happen to me of all people!'—'Oh! It's *very* pretty,' he said. 'It has my highest approval.' And he nodded approbation at the unclean car. Nothing could make him say that he couldn't see through the smoke and the layers of lithium powder.

His whole retinue stared and stared. One saw no more than another, but they all joined the intellectual in exclaiming, 'Oh! It's *very* pretty,' and they advised him to drive this perfectly clean car especially in the great procession he was soon to lead. 'Magnificent! Excellent! Unsurpassed!' were the words bandied from mouth to mouth, and everyone did his best to seem well pleased. The intellectual gave each of the swindlers a cross to wear in his buttonhole, and the title of 'Mr Clean Air'.

Before the procession the swindlers sat up all night and burned more than six canisters of oil, to show how busy they were finishing the intellectual's new car. They washed the car until it shone and pushed it out of the door of the plant with much huffing and puffing, and they swept up the lithium lying discarded all around. And at last they said, 'Now the intellectual's new car is ready for him.'

Then the intellectual himself came with his cleverest colleagues, and the swindlers each gestured towards the gleaming car. They said, 'See the charging cable, clean as clean could be, the oil-free moving parts, the power storage battery that lasts forever, the space-age design, and all as light as a spider's web. One would almost think there were a catch to this, but that's what makes the car so magnificent.'

'Exactly,' all the clever colleagues agreed, though all they could see was a large protrusion under the back seats housing an enormous lithium battery and dents in the floor where the car had recently stood.

'If our esteemed customer will condescend to take the wheel,' said the swindlers, 'we will help him with his first test drive.'

The intellectual stepped into the car and the swindlers got in beside and behind him. They showed him how to start the motor and they took off down the road, with the car running smoothly and quietly. The onlookers cheered.

All of a sudden, as the car was disappearing round a corner, there was a flash of orange and onlookers thought they saw flames shooting into the air. But the flames could not have been coming from the intellectual's clean, safe new car! 'There's an old diesel car parked down there', they said. 'It must be the diesel car that has burst into flames.' No-one had ever seen a diesel car burst into flames, but that had to be the explanation: diesel, after all, was a dangerous and obsolete fuel. Everyone knew that. Nobody would say otherwise, for that would prove him either unfit for his position, or a fool. No car anyone had driven before had ever been such a complete success as this magnificent new invention.

'But it's the new car that's burning!' a little child said.

'Did you ever hear such innocent prattle?' said its father.

But one person whispered to another what the child had said: 'It's the new car that's burning. A child says it's the new car that's burning.'

'But it's the new car that's burning!' the whole town cried out at last.

The intellectual's colleagues shivered, for they suspected that the people were right. But they thought, 'These cars must be the future, because the media say so.' So they patted each other on the back more proudly than ever, and made plans to buy their own electric cars.

The atmosphere has no roof and so there is a continuous flow of energy (and warmth) from and into outer space. The earth is no greenhouse.

—Bernard Jarman

VI.

What is causing climate change?

By Bernard Jarman

> Bernard Jarman has been involved with biodynamic agriculture most of his life. He managed a small biodynamic farm in North Yorkshire for a number of years and is currently estate manager at Hawkwood College near Stroud, Gloucestershire. He served on the board of the UK Biodynamic Association from 1995 to 2010 (5 years as chairman, 10 years as director). He co-founded Stroud Community Agriculture and later Common Soil Ltd (two CSA projects near Stroud) and actively supports independently inspired local initiatives. He also runs introductory courses in biodynamic agriculture, is fluent in German, translates texts into English and regularly works as an interpreter.

Everyone today is talking about climate change and the need to curb ever increasing CO_2 emissions. The very welcome consequence of this is that increasing numbers of people are now trying to reduce their environmental footprint and pursue a more sustainable lifestyle. There is also growing recognition that our pursuit of limitless consumerism is very destructive. Human actions are changing global climate patterns and the weather has undoubtedly become more unpredictable, yet is it right to assume that the emission of CO_2 is the primary cause?

The CO_2 global warming hypothesis originated in the research carried out by Swedish scientist Svante Arrhenius. His field of research involved chemistry and physics and one of his key discoveries was that mineral salts dissociate into pairs of ionized particles when dissolved. He also sought to develop a theory that could explain the ice ages. For this he drew on the work of John Tyndall (1820–1893) a physicist who studied radiant heat[1] and demonstrated how certain gases could absorb this heat. He found water vapour is the strongest absorber of heat and that certain compound gases notably CO_2, ozone and methane, absorb heat as well, though to a far lesser extent.

Arrhenius was able to show that an increase in atmospheric CO_2 concentrations will result in more radiant heat being absorbed. In 1896 he concluded that 'if the quantity of carbonic acid increases in geometric progression, the augmentation of the temperature will increase nearly in arithmetic progression.' Out of this he developed the hypothesis that increased concentrations of CO_2 caused by the burning of fossil fuels, will warm the planet. Since that time many scientists have verified the heat absorption capacity of CO_2 and as a result the contention that increasing CO_2 levels have a warming effect on the planet, appears indisputable. It is worth remembering however that main focus of Arrhenius' work was on radiant heat and in his calculations he consciously omitted to include the effect of clouds and heat convection.

Heat always has the tendency to move from something warm to something cooler and is transmitted in three different ways—conduction (the transfer of heat through solid ma-

---

1 Radiant heat originates primarily from the sun. It is transmitted through the air or a vacuum and then absorbed to a greater or lesser extent by the solid and liquid materials of the earth. Some is then radiated back into the atmosphere and absorbed by water vapour and other gases, retaining warmth in the atmosphere.

terials), radiation and convection. As regards atmospheric warming conduction doesn't apply and radiation makes up only 8% of the total heat transmission in the atmosphere.

The main way that heat is transferred through liquids and gases is via convection. As soon as it is warmed by the heat radiating from the sun, other heat sources or heat reflected from the ground, the air begins to rise. Warm air rises and as it does so it gradually cools down. Cool air holds less moisture and so clouds form. In cooling down the warmth in the air is released and escapes into space. Once it has cooled the air descends again to complete the cycle. These convection currents cause the air to circulate round the planet. Hot air from the tropics rises and then descends as cold air at the poles. The rising warm air releases rain, the sinking cold air brings high pressure and blue skies. It is the intricate and ever changing relationship between warm and cold air masses that brings about our weather. These in turn are affected by the complex system of ocean currents in the hydrosphere and up above by the fast moving currents of the jet stream.

In recent research carried out in 2008[2] it is stated that: 'The convective component of heat transfer dominates in the troposphere. When infrared radiation is absorbed by the greenhouse gases, the radiation energy is transformed into the oscillations of gas molecules, i.e., in heating of the exposed volume of gaseous mixture. Then the further heat transfer can occur either due to diffusion or by convective transfer of expanded volumes of gas. Inasmuch as the specific heats of air are very small (about $5:3 \times 10-5$ cal/s/cm ˚C), the rates of heat transfer by diffusion do not exceed several cm/s,

---

2 G V Chilingar and L F Khilyuk of University of S California and OG Sorokhtin of Russian Academy of Science, Moscow, 'Cooling of atmosphere due to CO_2 emission', 29 February 2008, *Taylor & Francis* [website], <https://tandfonline.com/doi/full/10.1080/15567030701568727>, accessed 21 August 2024.

whereas the rates of heat transfer by convection in the troposphere can reach many meters per second. An analogous situation occurs upon heating of air as a result of water vapour condensation: the rates of convective transfer of heated volumes of air in the troposphere are many orders of magnitude higher than the rates of heat transfer by diffusion.'

Without an atmosphere the earth would either become unbearably hot or unbearably cold. It is the air and the convection currents moving through it that makes our planet habitable. It is also what ensouls and gives life to our earth. The atmosphere has no roof and so there is a continuous flow of energy (and warmth) from and into outer space. The earth is no greenhouse.

The earth is a whole living organism so what happens in one part of the world will inevitably have an effect on the whole. If even the flapping of a butterfly's wings in some distant place can cause a storm[3], then larger changes will surely do so too. Today the changes are not just natural but have been brought about by human activity. Through our highly advanced technology it has been possible to wreak changes that no previous generation could ever achieve. Today the very fabric of life is threatened by pollution, plastic waste, radioactivity, fracking and the imminent introduction of full spectrum coverage by the proposed 5G network[4]. All these things are affecting the climate but perhaps the most far-reaching threat to climate stability is deforestation. Our forests are not for nothing known as the lungs of the world. They not only absorb carbon dioxide and release oxygen they are also able to regulate the amount of rainfall an area re-

---

3   Edward Lorenz (1917–2008) meteorologist and father of Chaos Theory discovered that tiniest of changes could have a huge effect.

4   G Millar & Y Roux, '...and not only the bees', 2020, *Star & Furrow* [website], <https://biodynamic.org.uk/wp-content/uploads/2020/04/Star-and-Furrow-Spring-2020-no-133-33-36-%E2%80%A6and-not-only-the-bees.pdf>, accessed 29 August 2024.

ceives and moderate extreme conditions of drought or flooding—quite apart from their supreme ecological value.

According to the Tree Foundation, 51% of the earth's original forest cover has been lost. Loss of tree cover has a direct effect on the local climate and a large forest such as the Amazon rain forest produces 50% of its own rainfall through transpiration. It can therefore be safely assumed that a further loss of global forest cover will have an increasingly significant effect on the global climate. Indeed, according to a 2005 study undertaken by NASA 'Deforestation in the Amazon region of South America (Amazonia) influences rainfall from Mexico to Texas and in the Gulf of Mexico. Similarly, deforesting lands in Central Africa affects precipitation in the upper and lower US Midwest, while deforestation in Southeast Asia was found to alter rainfall in China and the Balkan Peninsula. It is important to note that such changes primarily occur in certain seasons and that the combination of deforestation in these areas enhances rain in one region while reducing it in another.'[5] This clearly shows that by removing vast areas of forest, the living rhythmic system of the earth (its natural convection currents and air circulation), becomes chaotic and unpredictable.

It also goes on to point out that 'Deforestation does not appear to modify the global average of precipitation, but it changes precipitation patterns and distributions by affecting the amount of both sensible heat and that which is released into the atmosphere when water vapour condenses, called latent heat,' 'Associated changes in air pressure distribution shift the typical global circulation patterns, sending storm systems off their typical paths.'

---

5   Goddard Space Flight Center, 'Tropical deforestation affects rainfall in North America', 20 September 2005, *Mongabay—Conservation News* [website], <https://news.mongabay.com/2005/09/tropical-deforestation-affects-rainfall-in-north-america>, accessed 21 August 2024.

The rapid loss of tropical forest—currently an area the size of the UK is destroyed each year—is thus arguably the greatest driver of climate change. Forest degradation is also occurring in temperate and polar regions too. Deforestation in the world's mountain regions results in massive flooding downstream and we may well ask what climatic consequences the destruction of Boreal forest in Canada and Siberia is having.

But there are of course many other issues that we need to address if we are to secure a healthy and balanced world ecosystem. Biodiversity loss in many regions has become acute— not because of global warming—but because of toxins in the air and water, the ruthless extraction of resources and a system of industrial agriculture that seeks to dominate rather than work with nature. Monotonous and near lifeless landscapes devoted to mono-cultures and vast-scale production render landscapes incapable of ameliorating the climate. The elemental forces of wind, water, snow and fire are then no longer tamed, but take on gigantic form and bear with them enormous powers of destruction when they are released—fire storms in Australia, hurricanes in the US and elsewhere huge floods.

Something needs to change in the way we approach the earth. We need urgently to rethink our economic system. Our western growth economy only succeeds by exploiting the poorest in society, the developing world and the planet. The earth is still considered a resource to be exploited instead of a living being that needs caring for and managing. If we had a system whereby every resource that is mined or manufactured is returned to the earth at the end of its useful life in a form that allows for its re-incorporation into the natural order without toxins and residues, it would become sustainable but our capitalistic system could not function. We need a new system that is based on mutual service instead of personal profit. Only then can we meet the huge task that is facing us. The indigenous peoples of the earth have long known how to manage fire and water in the landscape and use them carefully to support their production systems. Today however we need to manage not only the untamed earth but also the land which has been cultivated, exploited, industrialised and effectively ruined. This we can only do from a spiritual vantage point.

The climate crisis we are facing will not be solved by focusing on the reduction of carbon emissions. Carbon is an essential building block of life and indeed there is much evidence to show that increased carbon dioxide in the air improves plant growth. This does not of course alter the fact that as much carbon material as possible should be retained in the soil to build up its humus content and retain soil fertility. All organic residues that arise on the farm should be incorporated without wastage and leaching or dissipation as carbon dioxide—in other words the focus should be on enhancing the life and fertility of the soil by having a closed farm organism and a carefully carried out system of composting and soil management.

A key element of biodynamic agriculture involves the use of biodynamic sprays. Amongst other things they serve to activate life processes and bring them into movement. This means enhancing the engagement of plant and soil, stimulating the breathing in and out of carbon dioxide, oxygen and nitrogen, maintaining a dynamic equilibrium and keeping the earth alive. And this is what the sun is doing for our whole planet. Its warmth brings the air and water into movement and ensures that it becomes neither too warm nor too cold. But to be fully effective it also needs a healthy ecosystem. However, when the lungs of the earth are harmed and too much of forest is removed, the delicate and complex equilibrium of the earth's climate is disturbed and unpredictable weather events as we experience them today, become more frequent.

The Windfall Gains Tax (WGT) will be applied to land-holders in Victoria. This tax target is the increased value of a property when it is rezoned as a 'Windfall Gain'. This tax is designed to capture the value of windfall gains generated from government rezoning and redistribute it to the wider community. The tax will be calculated based on the increase in the land's value resulting from the rezoning.

—*Pathfinder Law*

VII.

A windfall

By Gail Foster

The following essay is an excerpt from the book Masks, Mandates and Mayhem—
How one family kept a sense of humour through the covid panic, chapter 23 and was
kindly offered to us by the author.

Yum, that chutney is good.' I closed my eyes and licked
the red, plummy, concoction from a spoon.

Ten chutney filled small glass jars stood upended on a
large breadboard. Blobs of crimson stickiness stuck to the
stone kitchen bench, a large steel pot, a wooden spoon and
measuring cup. Three large plastic bags, fat with plums sat on
the tiled floor.

'What's in it?' asked Sara as she flicked through a House
and Garden magazine.

'Plums.'

'I guessed that. The table is laden with plum jam. It's a bit
of a giveaway.'

'Sugar, vinegar, raisins, salt, onion, garlic, ginger, mus-
tard seeds and cayenne.'

'Mmmm ... what will you do with the rest?'

'I'll give some to you, some to my parents, dry some and
stew some. I will then give jam and chutney to you all and
maybe some jars for presents.'

'Wow, what a windfall,' said Sara.

'Did you hear about Victoria's Windfall Tax?' asked

Peter. He carried a batch of freshly cleaned jars into the kitchen and placed them near the sink.'

'Nothing would surprise me about the Victorian government.' I scrubbed at a stubborn, sticky, blob.

'The new windfall tax is all ready to go on July the first,' said Peter. 'If you are a hard working, forward thinking person and have set aside some land which could be rezoned and subdivided you may earn yourself some good money. If you think the Victorian government is going to let that happen you are sadly mistaken. If you earn $100,000 or more you will be taxed up to 50%.'

'I'm beginning to see who is having the windfall,' smirked Sara.

'They call it a value uplift.' Peter shook his head in exasperation.

'It sounds like a bra marketing program,' said Sara.

'Isn't lifting another word for theft?' I asked.

'Like shoplifting?' asked Sara.

'Exactly, only now they're lifting the whole country.'

'Plum jam, plum chutney, what will I make next?' I puzzled. 'I'm thinking stewed plums for the morning porridge.'

'Will you add some sweetener this time?' implored Peter. 'My mouth nearly lifted off last time. It certainly woke me up but not not in a good way.'

'OK, I'll add some dates.'

'I can see a marketing opportunity,' said Sara. 'Plum power, the breakfast you have when you need a lift!'

Peter looked unconvinced as he patted Murphy's head. Murphy looked up expectantly, ever alert to food possibilities. I moved the jam onto the table for labelling and cleared a space on the bench for a bag of plums.

'Why does the preserving need to be done in the heat of summer?' I pulled up my sleeves and wiped my hands on my calico apron. Sun streamed through the north facing windows.

'You would think dopey Dan would add a sweetener to his sour windfall tax but he has upped the land tax by as much as 19%.' Sara positioned Murphy on her lap and scratched his ears. Murphy half closed his eyes in bliss.

'And that's not all. He has also brought in a huge increase in stamp duty,' added Peter. 'Then there's the energy policies.'

'There's more?' asked Sara.

'If you have any money left you will be forced to use it to upgrade your property to meet the council's energy efficiency test. If you are unable to complete the work you are told to do, the council will do it for you.'

'What if you can't afford it?' I asked.

'Too bad, the property is transferred to the council.'

'You mean they get another windfall?' asked Sara.

'It sounds suspiciously like, 'You will own nothing and be happy' from the World Economic Forum,' I said. I placed a large pot of plums on the stove.

'It will be hard to own anything,' agreed Peter.

'Do not despair Peter, everything will be all right,' said Sara. 'I am reading here that Australia will be saved with a green economy. Renewable jobs, zero emissions ... blah, blah and a green steel industry, what ever that is. It does say however that the plan would require hundreds of millions, sorry, billions of dollars in investment.'

'I have a funny feeling the tax payer will be funding it and some big corporation is going to get the windfall.' I cleaned a tray on the dehydrator, added some sieved plums and turned it on.

'I'd keep preserving mum. We're getting the skinny end of the fat of the land.'

'You know, my friend Melanie, from Melbourne signed her email, 'Love Melanie, from Dictator Dan's Directorate.''

'It's a plum job if you can get it.' I took off my apron and poured a cup of tea.

Maps tend to give the illusion that cities and towns take up a great part of the country. However, this changes when you zoom into a map, i.e. as you get a closer view. There is a lot of green and human habitations are few, a very small percentage of the country.

Can you imagine that the entire global population would fit into the United Kingdom? The UK represents only 0.05% of the planet's surface and 0.16% of all land area. The number of people living on the planet is 8 billion. If the entire population of the world lived in the UK, every single person would have around 30 m² of land area to themselves, while over 99% of the Earth's land would remain uninhabited by humans.

Someone might object that there would be too little space to grow food. Then let's move to the USA with an area of 9,525,067 km². If everyone lived in the USA, then each person could have 1,171 m² of land, e.g. 271 m² for a residential home with 900 m² left for agriculture. The USA contains only 1.87% of the entire Earth's surface—or 6.40% of the Earth's land area. Therefore, let's spread the world's population across the globe... What's the result?

How can you call the Earth overpopulated when every human being could have a fair share of 18,319 m² (4.5 ac) land and 44,419 m² (10.9 ac) of water each!

—Joseph Lang

VIII.

The adventure of the Jack Russell

By Tinderella

Tinderella is a former accidental academic, now dabbling in satire and songwriting.

AN HOMAGE TO SHERLOCK HOLMES

BAKER STREET, LONDON

It was a freezing day in February and the clocks were striking fifteen. Comrade Starlin had secured our humiliating re-entry into the EU and continental time had been enforced. The Thames was frozen and the heat pump at 221B Baker Street emitted no warmth, what with there being no wind to turn the wind turbines which blighted the landscape. Holmes and I drank tea and smoked to give us a semblance of solace. My friend scraped at his fiddle in a vain attempt to improve the circulation in his fingers. There was a knock at the door and Mrs Hudson showed in our visitor, a short, plump chap with a dishevelled mop of blond hair. Holmes bade him sit in the armchair by the cacophonous 'heat' pump and he commenced to tell us his tale.

'I shall get straight to the point Mr Holmes. My Jack Russell, Doylum, is endless... on people's legs. That is to say, he is insatiable, even more lascivious than myself. It was bad enough when he displayed amorous intentions towards Mr Jabid's cavapoo, but much worse are the dry cleaning bills for the laundering of my stained suits. In short, the dog is a darned nuisance and I want the irksome thing castrated and preferably out of the picture to boot. The problem is that my wife dotes on it and will not hear of its neutering and certainly not of a premature departure from its mortal coil.'

Holmes perused the matter, puffing heavily on his pipe; eventually the stimulating effect of the tobacco brought the light of a solution to his keen eyes.

'We have seen many peaks in our time; peak woke, peak cycle lane, peak vape shop... but none has been so pernicious as peak dog. They torment one's existence with their vexatious yapping and snapping, their droppings... I could go on. My good friend Dr Watson would be delighted to relate the terrible tale of "The Hound of the Baskervilles" if only you had six hours to spare. The working dog is admirable, such as Toby the tracker mongrel in "The Sign of the Four", but the pet mania has gone too far. To surmise, a reduction in their numbers would be desirable. Yet we must tread cautiously. There are matters of law to consider, to wit the European Convention on Canine Rights. I shan't bore you with the voluminous detail, but suffice to say, bumping off dogs is out of the question.

'So, how to proceed? I suggest a more subtle approach. We could of course kidnap the beast and release it in some desolate spot in the North, where it would doubtless be taken in by the common folk and find gainful employment as a rat catcher. However, I prefer a more elegant and satisfying course of action. Yes, we shall perform "the switch".'

'I tell you it can't be done,' the blonde blusterer blurted out. 'It's madness, it's moonshine! Do you not think I have spent eons plotting how to replace the infernal pest with a sensible hound such as a bulldog? I fancy I should call him

Dodo, as Winston called his own pet mutt. But I digress; the switch is a non-starter.'

Holmes gazed morosely at the smart meter as his funds disappeared in the running of the hopeless heat pump. He countered, 'I approve of your choice of name for the dog; it will serve as a most appropriate allusion to the deserved and long overdue extinction of the fraudulent 'Conservative' Party. But I digress; it shall be the switch or nothing.'

With a shrug of his shoulders the blabbering blatherer bleated, 'Do what you must sir, I am in your hands.' Holmes bade me shut the door behind our departing visitor and moved swiftly to the basin, where he proceeded to wash his hands in a most meticulous manner.

That night I slept fitfully; my bed was frozen and icicles formed on my runny nose. My belly craved the animal protein it had so long been starved of by the plant-based dietary directives of the WHO and the WEF. I dreamed of kippers and poached eggs. Yet when Holmes and I sat down to breakfast, Mrs Hudson served us our daily ration of cold soybeans.

'My good Watson,' said Holmes, 'would you be so kind as to consult the Bradshaw? Today we shall venture to the coast.'

Day trip to Brighton

We took possession of the Jack Russell by the ruse of Holmes masquerading as a mobile vet, summoned by our client on some spurious grounds. Though it was bitterly cold as we hastened along the platform to catch our train, Doylum retained his virility and became romantic with my leg. I put this down to his fresh meat ration, mandated for the pets of politicians in the aforementioned Convention. I was sorely tempted to eat some of said ration myself, however we were constantly monitored by video surveillance cameras and the penalty of five years hard labour enforceable by local council officials dissuaded me.

As the train trundled past roads strewn with abandoned electric vehicles which had broken down in the cold, my friend told me of his plan.

'It has come to my attention that a Mr Gizzard of Brighton has a fine hound that will suit our purposes admirably. Admittedly it is not the specified bulldog. It is, however, a poodle, another breed favoured by Mr Churchill, which is sure to appeal to our client's tastes. Being a standard sized poodle rather than Churchill's preferred miniature variety presents something of a height problem in relation to the Jack Russell, but I am up to the challenge. Gizzard's creature is named Doofus, so when our client inevitably decides to rename it Rufus, as per Winston's pet, the beast will soon recognise it is he who is being addressed. The poodle currently has its coat dyed pink, which will work to our advantage.'

'But Holmes, surely both this Gizzard chap and the wife of our client will notice that a substitution has occurred, particularly due to the pronounced difference in stature of the dogs?'

'Fear not Watson, the Brighton fellow is so dim that he will not notice the difference. When we disembark from the train we shall have the coat of our current charge dyed pink at a dog grooming salon. Thenceforth we shall proceed to the Gizzard residence and deftly execute the switch, so that the new owner of the Jack Russell will be none the wiser.'

'But what of our client's spouse, will she not notice the switch?'

'Here's the genius of it Watson. With our quarry in our possession, we shall seek out a poodle parlour and have the formerly pink poodle dyed a more natural off-white colour. On delivering the pooch to our client, the increase in its dimensions will be explained to his wife by my professional diagnosis as a 'vet' of a sudden growth spurt caused by climate change. I hear that she is afflicted with Net Zero hysteria, so she is sure to accept my preposterous prognosis.'

BACK IN LONDON

Naturally, as with all my friend's cunning schemes, everything came to pass exactly as he had foreseen. The next morning's post brought a letter embossed with a muddy paw print. Holmes read it aloud. 'Well, well, our client professes himself most pleased with our efforts and has deposited the handsome sum of €5,000 into my Social Credit account. He mentions in passing that his wife is perplexed as to why he has taken to addressing 'Doylum' as 'Rufus', however he is confident she will get used to it as just another of his eccentricities. My dear Watson, we shall celebrate this evening by dining on steak and a bottle of wine. With some good fortune, there may be a little of the fee remaining so that we may breakfast tomorrow on kippers and eggs.'

Net zero? I am led to understand that all life on this planet depends upon CO_2.

—*Glastian*

IX.

Architects of their own demise

By Glastian

> Glastian's boyhood was mainly spent in the woods and coun-
> tryside surrounding the small mining village where he was
> born; this has instilled in him a love and respect of nature and
> all living things, with the understanding that our very exist-
> ence depends upon living in balance with nature, unlike the
> unnatural artificial world being imposed upon us via 'pandem-
> ics', 'climate change', etc. He has spent half his working life in
> the offshore oil and gas industry; this gave him an insight into
> the workings of corporate business and helped open his eyes to
> the unbelievable corruption that exists in the world today. He
> has been retired for several years now, and is committed to a
> better world, free from greedy, psychopathic billionaires.

Many years before the so-called pandemic, my wife and
I occasionally hosted friends for dinner, and vice versa
(doesn't happen much now, for reasons we all know). Very so-
ciable, with after-meal drinks and conversations and discussions going on sometimes into the early hours.

One particular evening, the topic of conversation dealt
with the matter of where we think we came from, life itself
and the planets. Our hostess (let's call her Mary, not her real
name), an intelligent woman, not afraid to fight her corner,
was really enjoying putting forward her views, views which I
could not agree with. So I asked Mary a question; touching

the coffee table, I asked her what it was. Of course she said 'a coffee table', to which I agreed but I then said it is also 'mass', i.e. something solid. I then referred to other nearby articles, including our persons, moving on to cars, houses, the Earth itself, the Moon, the planets in the Universe, all agreed as 'mass'. A now impatient Mary asks where am I going with this? So I ask her what is between Earth, the Moon and all the other planets, to which she replied, well it's space isn't it. And I say would you agree that space might be described as a void, no atmosphere, in other words 'nothing'? I suppose so says Mary, so what? Now I ask Mary to imagine that there are no planets, no stars, no Universe, just 'nothing'; but the Universe is there, we are here, so what Power put it there, more to the point why is it there? I'm afraid Mary's reaction was not what I expected; clearly upset, she said she didn't want to go there, it was too mind-boggling to contemplate. End of subject discussion for that evening!

So, regardless of how the Universe was formed, let's have a look at our Solar System. (I would state now that what I write is how I see things, not necessarily how they are). A simple view, the Sun in the centre, then Mercury orbiting nearest the Sun, then Venus, then Earth, Mars, etc. at increasing distances from the Sun.

But planets nearer the Sun are warmer than planets further away, Mercury> Venus> Earth> etc. Simply put, the further away from the fire you are, the cooler you get.

Now our planet Earth (and I stress again this is only my take on things) would appear to occupy a unique place in the great scheme of things. 93,000,000 miles away from the Sun, let's say for ease of calculation that the temperature at our equator is 40 degrees Celsius and at the poles zero degrees Celsius. The diameter of Earth is some 25,000 miles. Applying simple geometry, the radius of Earth is some 4,000 miles; so it could be argued that, at 93,000,000 miles from the Sun the

Earth's temperature is 40 degrees, yet at 93,004,000 miles it is zero. A variation of 40 degrees over 4,000 miles? Too simplistic perhaps, but let's run with this theory for a minute.

The rotating Earth orbits the Sun, taking 365 days to complete an orbit. But who's to say that the orbit is a perfect circle? What if the Earth's orbit varies in distance from the Sun, sometimes nearer, sometimes further away? Our theory would seem to indicate a change of 1 degree Celsius for every 100 miles Earth is closer to/ further away from the Sun. Again, simply put, nearer the fire, warm; further away, cool. Might this explain ice ages in previous years?

I would use the same argument as regards the present so-called global warming, except that records appear to show that thousands of years ago, the Earth was actually 2 degrees Celsius warmer than it is now. This is open to discussion, of course, but the fact remains that Earth's temperature has fluctuated greatly over the centuries and very little of it has been down to human influence. All down to natural occurrences in my humble opinion, all beyond our control. All governed by whatever great power created the Universe—until now?

So here we are in 2024, with the world's human population being sold a pup by a relatively few psychopathic megalomaniacs intent on having the planet for themselves, using whatever unnatural means they have to steamroller their crazy agenda; ably assisted by bought and paid for governments who don't care about you or me. Everything they do is against nature; chemtrails, 5G, 'vaccines', 15-minute cities, forced agriculture, food shortages, Net Zero, etc. In fact doing the opposite of what is really needed, insidiously poisoning the very means of our existence.

Net Zero? I am led to understand that all life on this planet depends upon CO_2. I'm no expert on the matter, but I'm pretty sure if any of these Net Zero pushers were asked what percentage of CO_2 is presently in our atmosphere, they wouldn't have a clue. I believe that CO_2 forms around 0.04% of the air we breathe, air which is mainly made up of approximately 79% nitrogen and 21% oxygen. (Incidentally, if our air's O_2 level fell below 19% or rose above 23%, we would be in serious trouble). Again, like Earth's unique position in our solar system, who, or what determined such a mix of nitrogen and oxygen was required for our existence?

Surely then, it must follow that these inhumane cretins playing God with a planet we all depend upon can only be described as idiotic architects of their own demise? Pity is, they'll take us and all life forms with them if we don't put a stop to them soon.

We were given a beautiful gift of a planet; please don't let them kill our golden goose.

They have campaigned against me for exposing the Nut Zero scam for exactly what it is: The most transparent attempt to impoverish a population completely unnecessarily.

—Dan Wootton, Reporter on Outspoken

X.

The government's Absolute Zero is absolute madness

By Veronica Finch

> Veronica is co-founder of The White Rose UK, which is part of the freedom movement that came to existence in 2020. She is author of *The White Rose—Defending Freedom* and *Hope Amidst a Tsunami of Evil*. She also writes editor notes for the monthly print *Freedom Magazine*.

Absolute Zero is the name of a 60-page document published by UK FIRES, a research programme sponsored by the UK Government[1]. The document is filled with strategies concerning how to reduce emissions to zero by 2050. When I go through this paper it feels like a long *first of April joke*, only it's not. The government really has a plan. The plan appears to be quite mad, but also—when taking a closer look—rather evil.

This is their key message, stated on page 5: 'The big actions are: travel less distance by train or in small (or full) electric cars and stop flying; use the heating less and electrify the boiler when next upgrading; lobby for construction with half the material for twice as long; stop eating beef and lamb.'

---

1 UK FIRES, 'Absolute Zero—Delivering the UK's climate change commitment with incremental changes to today's technologies', *UK FIRES* [website], <https://ukfires.org/wp-content/uploads/2019/11/Absolute-Zero-online.pdf>, accessed 29 August 2024.

Under the guise of protecting the planet the government has sprawled out a plan which, when observed closely, is something like a climate lockdown and contains massive restrictions on people's freedom, to the extent that travelling would become almost impossible, and healthy nutritious meat would vanish from the table. There will be restrictions on your private use of energy, how you heat your home, what building material you may use and what food you may eat.

Imagine for a moment, that 'saving the planet' wasn't the true reason for Net Zero (because it's not)—how would you feel if someone told you how much water or energy you're allowed to use in your home? What kind of food you should eat and what not? Where you're allowed to travel to, or rather that you're not allowed to travel anywhere further than 15 minutes within your town/ ghetto (you may not be able to travel anyway, because you couldn't afford it)? What would you call this? Net Zero or Absolute Zero is an absolute overbearing tyrannical plan that outdoes communist policies (which were and still are bad enough).

In the introduction, the document mentions: 'We have to cut our greenhouse gas emissions to zero by 2050: that's what climate scientists tell us, it's what social protesters are asking for and it's now the law in the UK.'

Similar to the pseudo-pandemic, the government plays a game by mentioning 'the scientists', to make sure that people blindly trust, because after all, the 'scientists' *know* what they're talking about, and we have no clue—so lets leave it all to them. But what kind of scientists are they? Their own paid for puppets whose overlords are the WEF global criminals! There are no trustworthy scientists among the Net Zero promoters. Who are the protesters? These are groups organised and sponsored by the very people who are driving the climate change agenda. Those taking part in the protests who really believe in the nonsense are 'useful idiots' and are used as cannon fodder. They help spread climate fear and give the false impression that the concern for climate change is coming from grass root movements. These 'protesters' are a minuscule part of society. They do not represent the people.

On page 6 there's a colourful chart which shows what needs to be implemented between 2020 and 2029 and subsequently between 2030 and 2049. The ultimate goals for 2050 are the following:

➤ All new vehicles are electric with a reduced size of average 1000 kg, smaller cars [what about large families?]. Road use at 60% of 2020 levels, reduced travel distances [think: travel restrictions, 15-minute ghettos...]

➤ Electric trains will be the dominant mode of travel for people and freight over all significant distances.

➤ All airports closed down

➤ All shipping declined to zero

➤ Interior heat: heat pumps and energy retrofits for all buildings. Heating powered on for 60% of today's use.

➤ All appliances meet stringent efficiency standards to use 60% of today's energy. Electrification of all appliances, no gas cookers, smaller fridges, freezers and washing machines [again, what about large families? Will we have a Chinese-style one/ two-child policy?]

➤ No more beef and lamb [to make the general public physically weaker?], no imports not transported by train, fertiliser use greatly reduced.

➤ No iron ore or limestone, demand for scrap steel and ores for electrification much higher.

➤ Cement and new steel phased out along with emitting plastics.

➤ Conventional mortar and concrete phased out, new builds have to be highly optimised for material-saving.

➤ Goods made with 50% of material.

➤ Electricity: four-fold increase [just where will all that

electricity come from?!], end of all non-electrical motors and heaters.

➤ End of fossil fuels [so that everyone is dependent on electricity and unable to store coal/ petrol, whilst electricity may become unavailable at times, or be arbitrarily switched off by companies?]

Do you agree to any of this? Were you ever asked for your consent? We were never asked because, naturally, no sane person would agree to this plan, unless, of course, they were brainwashed into believing the propaganda or bribed with heaps of money.

These decisions have been mapped out without our consent, they do not make sense, they are not economical and not 'green', they are tyrannical, destructive and dangerous. From 2050 certain restrictions on emissions will be lifted again. Something is very chilling about this. Do they reckon with having gotten rid of a great part of the population in the meantime?

By the time we have reached 2050, a lot of the politicians and globalists who now promote Net Zero will not be alive. Do they really care about the future of the planet or is Net Zero just an excuse for something else, something more sinister?

Bottom line: Net Zero will be used for gaining control over the population and for reducing population numbers en route towards an alleged goal of achieving zero emissions by 2050. Net Zero is a madhouse plan that will clearly fail. But will it nevertheless be imposed through control and force, causing grave poverty and excessive deaths?

Are you aware that CO_2 takes up less than one half of one percent of the atmosphere? It is a truly tiny proportion and at that level cannot possibly be any sort of threat to life on earth. What is the point of trying to reduce such a minuscule impact? Say we were able to reduce it to 0.03%, would that make any difference? The notion is simply bonkers.

At 78% nitrogen is the most predominant gas in the atmosphere and yet we are being told that it, too, is dangerous and must be reduced. Again, say we were able to reduce that to 77%, what impact would doing that have?

These two statistics show just how much the climate change scientists have fooled themselves, and drawn the world's political class into believing in their madness; forcing us all to pay for their Net Zero nonsense.

—*Hugh Williams, Managing Director, St Edward's Press Ltd*

XI.

UK Net Zero: A layman's look at its contribution to global warming

By Jeremy Wraith

INTRODUCTION

Carbon dioxide, (CO_2) is a trace gas, currently accounting for about 420 parts/million (ppm) or 0.04% of the atmosphere. It is an essential part of our life, as if it falls below about 150 ppm all vegetation will die and all life on earth with it. (See *Inconvenient Facts* by Gregory Wrightstone.) Satellite images have shown that higher levels of CO_2 have increased global greening, which increases life preserving global oxygen levels. Commercial growers also pump CO_2 into their greenhouses to vastly increase plant growth.

What is not generally disputed is the fact that CO_2 is also a 'greenhouse' gas, which means that it does affect the earth's global temperature by warming it. The most effective greenhouse gas is water vapour which occurs naturally in the atmosphere and over which man has no control whatsoever.

Another undisputed property of CO_2 is that its effectiveness as a greenhouse gas reduces logarithmically as the concentration increases. However, the rate at which its effectiveness reduces is not universally agreed. So two sets of data are produced below for comparison purposes.

GLOBAL WARMING

Figure 1 is copied from a lecture given by Dr Tom Sheahan[1]. The figure, compiled by Prof Happer, and Dr van Wijngaarden, clearly defines the effect on global warming due to increasing levels of CO_2. This shows that increasing levels of CO_2 from the pre-industrial level of 280 ppm (parts per million) to the warming effect at today's level of about 420 ppm is practically indiscernible.

Figure 1: H&vW graph

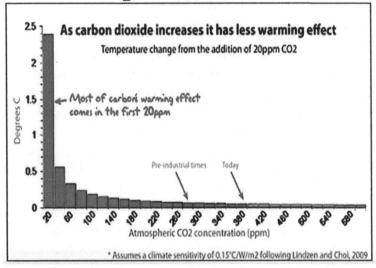

Figure 2, below is based on IPCC published information defining the effect on global temperature with increasing CO_2. This graph is copied from Gregory Wrightstone's excellent book, *Inconvenient Facts—The science Al Gore does not want you to know.* It also confirms the shape of the graph above.

---

1   YouTube [website], <https://www.youtube.com/watch?v=CqWv26PXqz0>, accessed 31 August 2024.

Figure 2, IPCC graph

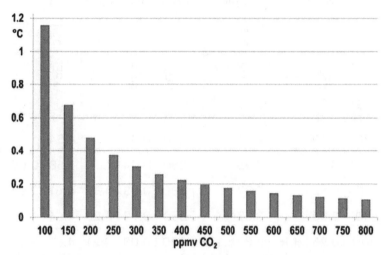

(Graph calculated using IPCC's formula $\Delta T_0 = \frac{5.35}{3.2}\ln\frac{C}{C_0}$;AR3, Ch. 6.1. Courtesy Monckton 2017)

The implications from Figures 1 and 2 are presented in Table 1 below, which shows the total temperature change (ΔT) as CO_2 rises from zero to 800 ppm. (NB The results have been scaled from Figures 1 and 2, so should be regarded as estimates rather than totally accurate values. However, the trend is clear regardless of the values presented. See the Appendix for additional information justifying Figure 1 and for ignoring Figure 2.)

Table 1: Temp. rise ΔT against CO_2 ppm increase: from figures 1 and 2

ppm	ΔT HAPPER	IPCC ΔT	ΔTHAPPER total	IPCC total ΔT	ppm CO_2	ΔT HAPPER	IPCC ΔT	ΔTHAPPER total	IPCC total ΔT
20	2.40		2.40		260	0.05		4.54	
40	0.55		2.95		280	0.04		4.58	0.00
60	0.32		3.27		300	0.03	0.30	4.61	2.94
80	0.30		3.57		350	0.03	0.25	4.64	3.19
100	0.25	1.16	3.82	1.16	400	0.03	0.21	4.67	3.40
120	0.15		3.97		420	0.03		4.70	0.00
140	0.12		4.09		450	0.03	0.20	4.73	3.60
150		0.65	0.00	1.81	500	0.03	0.17	4.76	3.77
160	0.10		4.19		550	0.03	0.15	4.79	3.92
180	0.09		4.28		600	0.03	0.14	4.82	4.06
200	0.08	0.48	4.36	2.29	650		0.13		4.19
220	0.07		4.43		700		0.12		4.31
240	0.06		4.49		750		0.11		4.42
250		0.35		2.64	800		0.10		4.52

The main points to note are:

1. The absolute minimum level of 150 ppm required for all vegetation and therefore all life on earth means a temperature increase, from zero ΔT, at zero CO_2, of 4.15 °C from H&vW and 1.81 °C from the IPCC.

2. The global temp increase at the 1830 pre-industrial level of about 280 ppm produced a total temp increase of

4.58 °C according to H&vW and a 2.85 °C increase according to the IPCC.

3. The current (say 2024) level is about 420 ppm, an increase of 140 ppm over 194 years, or 0.72 ppm/annum.

4. The total temp increase due to the rise in CO_2 to 420 ppm is approximately 4.7 °C (H&vW) and 3.5 °C (IPCC). This means that the global temp increase due to increasing CO_2 from 1830 to 2024 is 0.12 °C (H&vW) or 0.65 °C (IPCC).

5. But, the human contribution to this global increase, according to the IPCC, is 3% of the total. Hence, the current solution to the so-called global warming to reduce global human CO_2 to pre-industrial emissions will only reduce the global temperature increase by:

0.0036 °C (H&vW) or 3,600 ppm of 1 °C

or

0.02 °C (IPCC) or 20,000 ppm of 1 °C

6. As the UK only contributes 1% of the global human CO_2 this means that the UK will only reduce the global temperature after reducing its CO_2 output to pre-industrial levels by:

0.000036 °C (H&vW) or 36 ppm of 1 °C

or

0.0002 °C (IPCC) or 200 ppm of 1 °C

7. Also, it must be noted that increasing the current global CO_2 level from 420 ppm to 600 ppm is 4.82 – 4.7, (H&vW) and 4.06 – 3.5 °C, (IPCC) or 0.12 °C (H&vW) or 0.56 °C (IPCC).

8 Hence, increasing the global CO_2 by nearly 50% to 600 ppm from the current level of 420 ppm has a minimal effect on global warming. The global human contribution to that would only be:

0.0036 °C or 3,600 ppm of 1 °C (H&vW)

or

0.017 °C or 17,000 ppm of 1 °C (IPCC)

of which the UK contribution would be:

36 ppm of 1 °C (H&vW)

or

0.00017 °C or 170 ppm of 1 °C (IPCC)

9. Similarly, increasing the CO_2 level from the pre-industrial level to 50% more than the present level will *only* increase global temperature by

0.0072 °C or 720 ppm of 1 °C (H&vW)

or

0.0363 °C or 3,630 ppm of 1 °C (IPCC)

NB: If the global CO_2 level was doubled to 800 ppm the IPCC curve suggests that the temperature increase from pre-industrial levels will be 4.52 − 2.85 = an increase of 1.67 °C of which the human element would be 0.05 °C or 1/20th of 1 °C.

10. The total global rise in CO_2 from 1980, (335 ppm) to 2024 (420 ppm) was 85 ppm or nearly 2 ppm/annum over the last 44 years. Hence, it will take the earth nearly 90 years to increase the global CO_2 level to 600 ppm at that rate. This will only increase global temp by 0.12 °C (H&vW) or 0.56 °C (IPCC) at that level.

11. Assuming human emissions were 3% of the annual total of 2.02 ppm gives a global increase of 0.0606 ppm/annum. The UK share of that at 1% gives an annual UK emission figure of 0.000606 ppm/annum as the UK's increase in CO_2 over the last few years.

This means that it will take approximately 1,650 years for the UK to add just 1 ppm of CO_2 to the global total.

12. Hence, the world can carry on producing CO_2 at the current rate and after nearly 90 years it will still only have added 0.12 °C (H&vW) or 0.56 °C (IPCC) to the global temperature.

13. The results derived from the IPCC graph are considered to be unreliable and should therefore be ignored. There is ample evidence that the IPCC's reports and proced-

ures, and the activities of the global warming fraternity generally, are littered with examples of questionable practice. This includes many examples where they have ignored, manipulated and suppressed evidence that does not support their Net Zero agenda. Numerous publications, listed in the Appendix, describe in detail the many examples of their questionable practices.

COMMENTS

1. These figures demonstrate the total stupidity of the global Net Zero philosophy and it must be abolished at the earliest opportunity.

2. The current rise in global temperature of 4.7 °C, (H&vW), or 3.5 °C (IPCC), due to the current CO_2 level of 420 ppm has already happened and the world is still carrying on as normal.

3. Increasing global CO_2 level to 600 ppm will only add 0.12 °C (H&vW) or 0.56 °C (IPCC) to the global total and it will take nearly 90 years to reach that level at the current rate of increase.

4. Therefore, the current hysteria over the 'so called' effect of rising CO_2 levels causing disastrous increases in global warming, thereby causing melting of polar ice caps, more extreme weather conditions, etc. is entirely unnecessary.

5. Conversely, it should be noted the huge benefits to food production resulting from the increased CO_2 which promotes world plant growth and agriculture. Higher CO_2 concentration in the atmosphere increases food production and more life sustaining oxygen for all living creatures on earth.

CONCLUSIONS FOR THE UK

The misguided rush to reduce global warming by reducing CO_2 to pre-industrial levels is ruining the UK economy, its residents' livelihoods, living standards and freedom of movement.

In addition, the drive to Net Zero is totally unrealistic, totally unachievable and is going to cost the UK trillions of pounds to de-carbonise the grid together with all the other mandatory costs involved in diverting our manufacturing overseas.

For example, the drive to Net Zero has recently resulted in stopping steel production in the UK, mainly due to the closure of electric arc blast furnaces so as to reduce UK production of CO_2. Typical spurious virtue signalling, as more CO_2 will be generated by making the steel abroad and transporting it to the UK!

The imminent closure of the UK's one remaining coal powered generating station is another example. This is clearly ridiculous as 1,893 new coal powered generating stations are being built in the world. The total number in operation will then then increase from 3,743 to 5,636. Of these the EU has 465 existing plants and is adding 25 giving a total of 490 plants.

So, it is not really in the best interests of the UK to abolish steel making and throw thousands of skilled craftsmen out of work for the sake of trying to achieve Net Zero thereby saving 36 ppm of 1 °C.

It is also necessary to explain how the UK's 0.000165 °C (0.55 x 3% x1%) maximum extra contribution to global temperature over 194 years, or on average, 0.0000008 °C/annum, has endangered the earth so much that it justifies the Net Zero legislation, all the trauma that goes with it and unjustifiable reparations to third world countries.

APPENDIX—THE ACCURACY AND FEASIBILITY OF THE
H&vW AND IPCC GRAPHS

The two graphs presented in Figures 1 and 2 of the note are similar in shape but show different results. It is therefore necessary to examine which graph is more meaningful and accurate.

The Happer & van Wijngaarden results in Figure 1 can be justified by means of the following graphs which show excellent correlation with measured results:

The figure 3 below is also copied from the lecture given by Dr Tomas Sheahen[2].

Figure 3: 'Methane—The irrelevant green house gas', Dr Thomas P Sheahen

Stunning agreement with measurements

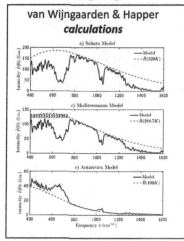
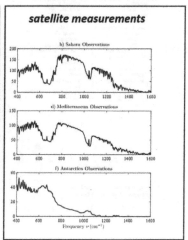

With regard to the IPCC results, Andrew Montford's excellent books *The Hockey Stick Illusion* and *Hiding the Decline*, which details the history of the 'Climategate Affair' show how the IPCC operates. These and other books (see list below) are essential reading to understand the workings and methods employed by the IPCC. These clearly show that the IPCC and the authors of IPCC reports are quite willing to edit information and ignore results that do not fit in with their intention to promote global warming at every oppor-

2 *YouTube* [website], <https://youtube.com/watch?v=CqWv26PXqz0>, accessed 14 September 2024.

tunity.

In addition, the graph below (also in David Craig's excellent book *There Is No Climate Crisis*) shows the results of IPCC estimates of global temperature increase over time This clearly shows the IPCC results are well over actual results.

Figure 4

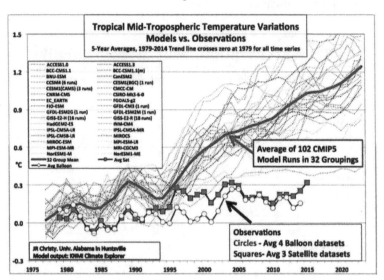

Hence, in view of the above and more evidence of IPCC failings to represent real values it can be assumed that the IPCC results are not reliable and should be ignored.

References

Christopher Booker, 'The Real Global Warming Disaster', Continuum, 2009

A.W. Montford, 'The Hockey Stick Illusion', Stacey International, 2010

A.W. Montford, 'Hiding the Decline', Anglosphere Books, 2012

Gregory Wrightstone, 'Inconvenient Facts', The science that

Al Gore doesn't want you to know, Silver Crown Productions Ltd., LLC, 2017

Bruce C Bunker, PhD, 'The Mythology Of Global Warming', Climate Change Fiction vs. Scientific Facts, Moonshine Cove Publishing LLC, 2018

M J Sangster, PhD, 'The Real Inconvenient Truth', Amazon, 2018

David Craig, 'There Is No Climate Crisis', Original Book Company, 2021

Ian Plimer, 'Green Murder', A life sentence of Net Zero With No Parole, Connor Court Pty Ltd., 2021

Dr Niall McCrae, RMN, MSc, PhD, 'Green in Tooth and Claw', The Misanthropic Mission of Climate Alarm, The Bruges Group, 2023

GMO crops—soybeans, corn, potatoes, canola, sugar beets, along with common fruits and vegetables—have infiltrated our food supply. They form the backbone of factory-manufactured, highly processed Frankenfoods marketed to consumers as wholesome and healthy while traditional foods—particularly of animal origin—are maligned.

—T. J. Martin

XII.

Double dare

By T. J. Martin

> To T.J. Martin language, literacy, thought, ethics, the environment, health, autonomy, freedom and sailing are all of great importance both in this essay and in life.

It made great theatre back in 2019—the saga of a heroic young climate warrior: her school boycott, her stoic mien, the triumphant Net Zero Atlantic sailing adventure, her pure outrage thundered in a single question echoed on the world stage:

'How dare you?'

I've since thought much about the footprint left in my wake. And although my orbit is indeed much smaller—and more mundane—I also have a question or two regarding this planet we call both Mother and home.

My first question, admittedly, is rather humble:

Do you have a clothesline or even a laundry drying rack?

Most people, I've noticed, do not air dry—even though utility companies extol pennies to be saved by conserving gas or electricity while simultaneously rescuing Mother Earth from her current distress.

Pennies saved are the least of my concern, however. Along with doing my planetary duty, I love the opportunity to spend time outside amongst the flowers and trees, in breeze and sun, enjoying the hopefully—fresh air.

The freshness of neighbourhood air is unfortunately con-tingent on which household is doing laundry at any particu-lar time. Those who use commercially marketed laundry products—products which promise to freshen clothes—suc-ceed quite well in dirtying the air we all must breathe. Unreg-ulated and undisclosed, the volatile organic compounds which give these products scent are heated and released with every dryer tumble, fouling not only the atmosphere, but our bodies as well.

How green is that? Why are these harmful chemicals not illegal—or at least disclosed and accurately labelled by those profiting from the sale of these toxins?

Another related question follows:

What fabrics do you wear and use each day?

Are your clothes, towels, sheets, blankets, comforters sourced from natural fibres like cotton, wool, linen, silk? Or do you wear and sleep on petroleum-extracted composites such as polyester, nylon, acrylic, spandex—materials which release multitudes of micro and nano toxic particles into the water supply with each washing?

Both environmental and health problems are caused by these synthetics which are chemically coaxed from oil. They do not proceed from nature, thus nature does not reabsorb them through decomposition; rather, they break down phys-ically into ever tinier shreds, pieces, strands, bits. Treated fabrics—rainwear, flame retardant upholstery, even children's pyjamas—are particularly dangerous as their coatings are composed of extremely stable molecular bonds. In living creatures, these forever chemicals disrupt the normal func-tioning of bodily hormones, leading to the recent phenomen-on of androgynous frogs and fishes which, after all, live in chemically tainted waters.

And herein lies the very big problem: No matter its size,

shape, or form, plastic is forever!

Barring a major scientific breakthrough, even the sleek carbon fibre yacht which carried our young climate warrior across the sea could, one day, itself float under and not on the waves, deteriorating into ever smaller chunks and particles, swirling about in a vast oceanic gyre of toxic plastic debris, eventually entering the marine and even the human food chains.

'To this end all plastics must come.'

So, if Big Oil is environmentally bad, why wear it every-day? Why drink from plastic? Why brush your teeth with it? Bag your groceries in it? Cook on it? Why is plastic not highly regulated and taxed as the toxin it is instead of manu-factured and sold to us in so many colourful forms?

Where is the outrage? The boycott? The revolt?

Thinking of Big Oil and its many hydra-like applications leads to yet another question:

Do you have a garden or make any effort to grow a bit of your own food organically?

Rather than relying on the global corporatocracy to clothe and feed you, have you ever watched a seed take root in the earth, sending delicate white hairs downward while reaching tiny green leaves toward the sun? Have you ever ex-perienced the later wonder in harvesting, say, a living tomato grown from that little seed?

Sadly, all too many of us have almost no idea from whence cometh our food. As with chemically sourced textiles and scents, our food supply has been severed from its roots and transformed into a commercial enterprise vaulting profit above people. Food is soullessly mass-produced on soils de-pleted of biological diversity, heavily sprayed, then harvested, processed, and transported vast distances to consumers.

Even more concerning, the very essence of life itself has been breached by new-age CRISPR technology capable of altering genetic codes which, in nature, evolved over eons of propagation. Cold-water flounder genes are shot onto the chromosomes of tomatoes to enhance shelf-life; eel genes have been inserted into Atlantic salmon to enhance growth; pesticide producing genes have been integrated within the genome of corn.

'O Brave New World!'

What could possibly go wrong?

Contrary to the official narrative supporting the genetic modification of our food supply, all is not GRAS, i.e. Generally Recognized As Safe, with GMO crops.

Vast amounts of petroleum-derived pesticides are necessary to grow these costly, one-off, genetically modified seeds. Whereas in nature, seeds genetically self-select from generation to generation, GMO seeds will not produce viable offspring beyond their initial planting. Yields do not rival those from traditionally grown crops; chemical pesticides are not spared; water is not conserved. The land is neither happy nor fruitful.

How green is that?

GMO crops—soybeans, corn, potatoes, canola, sugar beets, along with common fruits and vegetables—have infiltrated our food supply. They form the backbone of factory-manufactured, highly processed Frankenfoods marketed to consumers as wholesome and healthy while traditional foods—particularly of animal origin—are maligned.

Where is respect for the creation? Harmony with nature? Reverence for the old ways?

Meanwhile, the general health of the population continues to decline while maladies such as obesity, diabetes, heart disease, and cancer have reached levels unimaginable a few generations ago.

Speaking of past generations, few from them could even fathom the omnipresent technology permeating our lives today, which begs the next of my questions:

How invested are you in the cyber, digital, manipulated universe?

Could you survive without the constant companionship of multiple devices tethered to the Internet?

We have come to rely on the World Wide Web for everything, it seems. Information, opinion, news, entertainment, instruction, goods, services, friendships, relationships, community, self-esteem, world view, all flow from your ISP—for a monthly fee, of course.

Hardly a mention is made, however, of the massive environmental impact of this cyber infrastructure with its satellites, towers, servers, cables, hardware and software, and devices. All, of course, are manufactured from tangible resources—not to mention oil—extracted from the earth.

The current level of electricity needed to power this alternative universe is staggering. And, as small and large engines go green by plugging in rather than filling up, more and more power will be required.

Long term health effects from the electromagnetic fields generated by these towers, wires, and devices may take years to manifest, but the psychological ramifications of living in the cyber universe, sadly, are only too obvious amongst earth's population.

The cognitive disconnect, or dissonance, is astounding. On one hand, we are chided for changing Earth's climate through our evil consumptive ways, yet the same talking media heads report dispassionately of war and destruction without batting an eye at the environmental costs of such conflagration, human suffering notwithstanding.

Long before the Internet, the Native tribes of America's Pacific Northwest lived quite well despite their high population density. The annual salmon run provided an essential food source, so elaborate fishing rituals were in place between downriver and upriver tribes. Prescribed fishing days and quotas were enforced, especially by downriver tribes, allowing fish to swim upriver, thus ensuring a fair distribution of the resource and avoiding conflict. Their elders wisely understood what our current rulers do not—that war is a tremendous waste of precious resources.

So who are the powerful, the rulers of our day, those who believe themselves anointed both to instigate international conflict and simultaneously save the earth by convincing the rest of us that doomsday fast approaches?

Just who benefits from the climate change doomsday narrative?

Indeed, there seems to be a shadowy group of very rich elites who—along with earth's air, water, resources—deem humanity just another commodity to be managed and exploited, herded and controlled.

Do these privileged masters worry about their massive carbon footprint as they travel via private jets to idyllic locales, like Davos, to map out a sustainable future—for the rest of us?

Will they actually live in the 15-minute cities they envision, or will they be comfortably ensconced on vast holdings, feeding their cattle organically whilst we are advised to subsist on manufactured 'meat,' cricket meal and GMO crops in order to save Mother Earth?

Will they also 'own nothing and be happy'?

Or is such joy exclusively reserved—for the rest of us?

Are we to be rootless serfs, albeit 'happy' ones, in a new global feudalistic iteration?

Ironically, the good intentions of many potentially 'happy' serfs have been hijacked by those pulling the strings of the climate change narrative, by the rich and powerful few.

Whether through lack of critical thought, youthful idealism, or herd mentality—many well-meaning folk have been converted and are now avid proselytes themselves! These workaday warriors and do-gooders so blindly commit to saving Mother Earth from this fabricated existential threat that they completely ignore the environmental outrages committed each day against our air, water, food, health, and well-being.

They believe at the expense of seeing.

Climate hysteria thus becomes climate hypocrisy and its lowly disciples mere pawns in a global game of thrones.

Cui bono? To whose good?

Obviously, those who most passionately perpetuate the climate change myth stand to most benefit from it.

How dare we not question their dark motives of profit, power, control, and global hegemony?

How dare we not collectively demand they answer the most important question of all:

HOW DARE YOU?

The reality that CO2 has negligible impact on climate means that all our efforts to reduce CO2 emissions have been pointless.

—*Douglas Brodie*

XIII.

Debunking the climate change hoax, part two

By Douglas Brodie

The reality that CO_2 has negligible impact on climate means that all our efforts to reduce CO_2 emissions have been pointless. The deployment, against the advice of a government Chief Scientific Advisor, of intermittent expensively-subsidised so-called renewables like wind turbines and solar panels which end up as toxic non-recyclable junk when they reach the end of their short service lives should never have been embarked upon in the first place. This has been money straight down the drain, or rather money straight from the pockets of the general public into the coffers of Big Money. Equally misguided green hydrogen and 'bonkers' carbon capture and storage are certain to be ruinously expensive. Such inappropriate technologies and the legally-obligated push for Net Zero are condemning the country to long-term economic decline.

A few final bullet points to sum up the insanity/ malevolence of Net Zero. I have put such points to many parliamentary representatives over many years but they have always fobbed me off or not replied at all:

➤ Atmospheric CO_2 levels have been rising steeply for decades, uninfluenced by the various 'landmark' climate agreements (or Covid global lockdowns). There is no indication that decarbonisation efforts to date have had any effect on the rising trend or that intensified decarbonisation efforts could arrest it any time soon, far less turn it into steep decline.

➢ The UK contributes just 1% of global CO_2 emissions so irrespective of the disputed science of alleged man-made CO_2 global warming, attempting to decarbonise unilaterally is pointless given that the hugely more populated non-Western world is not going to follow suit any time soon. Why are we pointlessly committing economic suicide?

➢ The world as a whole is still around 84% dependent on fossil fuels, an unbridgeable chasm away from Net Zero. After 15 painful years of Climate Change Act/ Net Zero striving, the UK is still around 80% dependent on fossil fuels, clear proof that decarbonisation is going nowhere. The so-called global green energy transition is a going-nowhere fiction.

➢ The UK government was told in a recent report commissioned by the Department of Energy that it has no hope of reaching its Net Zero targets (which has been obvious for many years) but is ploughing ahead regardless, a clear indication that the real purpose of Net Zero is to drag us down into legally stitched-up deindustrialised immiseration.

➢ Trying to decarbonise the grid by relying on intermittent wind and solar power without the 24/7 balancing and back-up currently supplied by fossil fuels (the UK's last coal power station will be retired in October) will inevitably lead to prolonged blackouts and/ or severe energy rationing. The UK Energy Secretary has no idea how to avoid this self-imposed disaster judging by her shockingly naive reply to recent questioning. Battery storage costing multi-trillions is not an option. Allowing this disaster to unfold looks like a diabolical plan by the green blob to drag the economy down.

➢ Attempts to persuade or coerce the public to adopt unwanted and pointless EVs and heat pumps are going nowhere other than leading to, inter alia, the ruination the car industry as made clear in these articles[1]. Again, this looks like part of a diabolical plan to 'collapse industrialized civilizations', to paraphrase UN IPCC architect Maurice Strong.

CONCLUSIONS

I know that many people find the subject of 'climate change' too arcane and daunting to challenge and that they prefer to opt for the comfortable assumption that the establishment authorities must be working for our best interests. I hope this paper shows that, to the contrary, the establishment and their puppet politicians are intent on causing us serious harm and that the evidence for this is clear.

The UK government recently rejected a petition to repeal the Climate Change Act 2008 and the Net Zero targets. Compare the clear-cut evidence given above with their mendacious, boilerplate response, including their cheating claim of having halved UK CO_2 emissions when in the main they have only been offshored, at the cost of swathes of UK jobs. It

---

[1] —Paul Homewood, 'Law to limit petrol car sales is "terrible for the UK", warns Vauxhall maker', 26 April 2024, *Not a Lot of People Know That* [website], <https://notalotofpeopleknowthat.wordpress.com/2024/04/26/law-to-limit-petrol-car-sales-is-terrible-for-the-uk-warns-vauxhall-maker>, accessed 30 August 2024.
—Paul Homewood, 'Ford boss says it may restrict petrol models in the UK to hit EV targets', 8 May 2024, *Not a Lot of People Know That* [website], <https://notalotofpeopleknowthat.wordpress.com/2024/05/08/ford-boss-says-it-may-restrict-petrol-models-in-the-uk-to-hit-ev-targets>, accessed 30 August 2024.
—'The Used EV Timebomb', 1 May 2024, *Not a Lot of People Know That* [website], <https://notalotofpeopleknowthat.wordpress.com/2024/05/01/the-used-ev-timebomb>, accessed 30 August 2024.
—'Drivers may struggle to buy petrol cars as garages face risk of £15k net zero fines', 16 May 2024, *Not a Lot of People Know That* [website], <https://notalotofpeopleknowthat.wordpress.com/2024/05/16/drivers-may-struggle-to-buy-petrol-cars-as-garages-face-risk-of-15k-net-zero-fines>, accessed 30 August 2024.

is surely obvious that our politicians are lying and that their 'really very stupid' climate change narrative has nothing to do with climate. It's all about imposing deep state totalitarian control over the people and global resources, all covertly planned out many decades ago.

All of the main UK political parties, Conservative, Labour, Lib Dem and SNP, often referred to collectively as the Uniparty, firmly back the climate change hoax and together disenfranchise the electorate on this and other globalist impositions. They are wilfully leading us into already well advanced deindustrialisation, 'Absolute Zero' levels of privation and, without fossil fuels, pre-Industrial Revolution living conditions.

What a horrible mess! My forlorn suggestion for getting out of it is that people need to stop voting for the treasonous Uniparty. I am not advocating any particular challenger party but if the Uniparty were to get a very low number of votes they would have much reduced moral authority, which would at least be a start. If constituencies could somehow organise themselves to focus their votes on a single anti-Uniparty candidate they could avoid splitting their votes, a mimic in reverse of how the minority SNP separatists keep winning here in Scotland because the tribalist votes for the unionist Con, Lab and Lib parties get split.

So then, brethren, we are not the children of the bondwo-
man, but of the free: by the freedom wherewith Christ has
made us free.

—*Galatians 4:31*

XIV.

What climate crisis? It's a freedom crisis

By Janet Sugden

Janet was born in Oxford and moved to London in the 1980s to attend art college. She has lived and worked in London ever since. She works part-time as a nursery assistant and paints in her free time.

There isn't a climate crisis. We are being lied to again. There is though a freedom crisis. Our freedom is being taken away from us gradually and continuously. There are 15-minute cities which limit our freedom to travel other than limited distances. There are changes in law that limit our freedom of speech. Governments around the world have been infiltrated by the World Economic Forum and the main goal of the WEF is to have control over all people of the world and the tactics they are using are instilling fear, because fearful people are easier to control, and lying. We have been told so many lies over the past few years. Many people struggle to believe the breadth and depth of the evil that we are being subjected to.

There have always been changes in the climate, but there is no need to be alarmed that the climate is getting hotter and hotter and will reach unsurvivable boiling temperatures. It is just another lie that we are being told. The governments are lying to us. The mainstream media is being paid by billionaires to lie to us. Some celebrities are being paid to lie to us. Some well meaning people are repeating the lies as they believe it's the truth. We were lied to about Covid-19 and the need to take a 'vaccine'. It was a lie that they were 'safe and effective'. It was a lie that people could not catch Covid-19 if they had taken the jab and it was a lie that it could not be passed on by those that had taken the jab. These lies have now come to light. There are excess deaths and millions of vaccine injured people.

The lies are keeping people fearful and being in fear is a low vibrational state. I believe that the way we can get through this is to raise our own individual vibration through creativity—painting, music or creative writing or anything that makes us feel joyful. Individuals may feel that they have very little power to make changes against this tide of evil that is upon us, but if everyone does what they can collectively we can make a difference.

We are in a spiritual battle and the fear being stirred up over climate change is getting people to lose their focus on what is important in life. People are being deceived. Many millions were deceived about the mRNA jabs and now also the climate change hoax. There are many verses in the bible warning people not to be deceived. Deception will increase in the end times as satan will be more active. He is known as the great deceiver.

We are being told that the 15-minute cities or ghettos, as they will become, reduce your carbon footprint and thereby help in saving the planet. More and more restrictions will be placed on us if we comply—climate lockdowns, euthanasia, more surveillance cameras to track our every move and digital ID. All this to trick us into believing that it is people that are causing the change in the climate and that restricting people from travelling and ending people's lives will help to solve the problem, but there is no problem to solve. We are being enslaved and impoverished under the pretence that this is saving the planet. The planet does not need saving; we do.

And how do we save ourselves? We could pray to our Heavenly Father, Creator of Heaven and earth, to ask for assistance and ask Him to send his angels down to earth to fight this battle. Another lie that has been fed to us over the years is that God does not exist. The tactics used to separate us from our loving God have been mockery and ridicule. Maybe it's time to turn back to God and become the free, happy people that God created us to be.

Any small increase in atmospheric CO_2 is a consequence of warming and is not the cause of global warming.

—*Louis Brothnias*

XV.

Anthropogenic climate change— scientific fraud, part one

By Louis Brothnias

'*None are so hopelessly enslaved, as those who falsely believe they are free. The truth has been kept from the depth of their minds by masters who rule them with lies. They feed them on falsehoods till wrong looks like right in their eyes.*'—*Johann Wolfgang von Goethe*

'*No amount of evidence will ever persuade an idiot.*'—*Mark Twain*

The alleged climate change crisis being of man's own making is presented only as a one-sided argument. The public debate is heavily censored by the mainstream media (MSM). If it were to be properly debated with real science and genuine scientists then the fraud would very quickly unravel. What is a normal temperature? Dramatic figures like CO_2 levels increasing by 47% (280ppm to 412ppm) in the last 260yrs are egregiously misleading. The parts per million (ppm) is very conveniently ignored. Subjectively, 280ppm → 412ppm = 132ppm (47% increase). Objectively, 280ppm = 0.0280% and 412ppm = 0.0412% atmospheric CO_2 content, so 99.972% of the atmosphere content is initially not CO_2 and after 260 years becomes 99.9588% not CO_2. The increase is an additional 132ppm = 0.0132% over a 260 year period.

Subjectively = 47% and objectively = 0.0132%

Both terms are mathematically correct but there is a difference >3500 in the level of CO_2 depending on how figures are presented. To be persuaded that 412ppm (0.0412%) carbon dioxide (CO_2) could be responsible for a global 'meltdown' is to believe the logic of the environmental idiot. Higher CO_2 results in a greener planet and better crops (photosynthesis). Vegetable growers routinely increase the CO_2 content in a greenhouse environment to generate the glucose within the vegetable (O_2 is a bi-product) to provide nutrition and improve the quality and yield of their products. A cooler medium will have a higher concentration of a dissolved gas than a warmer one. Warming sea water releases dissolved carbon dioxide (submarine volcanos generate tremendous heat and copious volumes of CO_2 and this underwater source of CO_2 and heat is ignored). The climate change fraud claims that a greater concentration of CO_2 results in warming seas and is left unchallenged. As the medium warms, this gas is released, not dissolved so that the atmospheric concentration will naturally increase.

That carbon dioxide causes global warming and warming seas is the fraud. Any small increase in atmospheric CO_2 is a consequence of warming and is not the cause of global warming.

The issue is not that there is no change in climate but that the alleged warming is caused by the use of oil-based fuels. The false reasoning is that increased atmospheric CO_2 results in global warming and comes about directly from burning oil-based fuels. Oil and gas are not fossil fuels. They are formed from non-living (inorganic) matter by an abiogenic process that requires intense pressure and temperature as was postulated at Earth's formation. Not rotting trees or vegetation. Both oil and gas are found far deeper than any possible organic source. That humans are responsible for their own demise is predicted on deliberately misrepresented (absolutely wrong) theory. Lies.

When ice melts, there is no measurable volume change.

The density of ice is less than water (it floats) as it contains trapped air. About 10% of the ice is above water. When the ice melts, this air is released and the reduction in volume accounts for that extra volume due to the trapped air. The amount of water contained in an iceberg as solid ice is exactly the same as is released as liquid water. The promoted illusion is that the disappearance of an iceberg results in an increase in the water volume. This can easily be shown to be false by filling a glass that contains a cube of ice with water to the brim. When the ice (that rises and is seen partially above the surface) melts there is no overflow of water. The volume remains the same. Dramatically rising sea levels caused by melting icebergs is pure scaremongering and is ... fraudulent science.

The atmosphere gases trapped in ice:

Gas	Composition	PPM
Nitrogen	78.00%	780000
Oxygen	21.00%	210000
Argon	0.96%	9558
Carbon dioxide	0.04%	412
Neon	0.00%	18
Helium	0.00%	5
Methane	0.00%	1.8
Krypton	0.00%	1.1

The release of the heat (the exothermic latent heat of melting) into the atmosphere could (theoretically) introduce instability and potentially result in weather disturbances as the transfer of heat causes winds to be created. However, such release is very slow and any weather change would be minimal. Originally, there was no atmosphere and as a consequence the level of the sea would have been much higher than it is today. The atmosphere matured over hundred millions of years into its present-day composition and has an estimated weight of 5.15×10^{18}kg (5.15 quadrillion tonnes). This enormous weight presses down on the water surface forcing water underground to form subterranean lakes. The depression of sea level provides an explanation as to why chalk cliffs rise so high above the present-day sea surface. Chalk originates from the shells and skeletons of primitive water-borne creatures. Water vapour also entered the atmosphere to form clouds (the visible evidence of water) and these clouds as water in its condensed liquid form are blown by the wind and subsequently precipitate their load as rain, hail or snow somewhere else. Relentlessly, the cooling towers of power stations are shown with the obviously fraudulent intention that the nebulous visible clouds are carbon dioxide. Carbon dioxide is a colourless (invisible) trace gas. The reality is that condensing water vapour forms clouds of steam. The water cycle neither creates nor destroys water but simply redistributes it around the World. Humidity, moisture, and condensation demonstrate the presence of water and water vapour is a powerful greenhouse gas.

Consider: why it is extremely hot in the daytime and so cold at night in the Sahara and Gobi deserts? There is no moisture in the atmosphere.

Hiding true agenda starts with concealing or manipulating data. Using a name that hides your true intention or changing it to confuse people can delay getting caught.

Another topic which doesn't have anywhere near enough information or coverage is Solar Radiation Management as the government likes to call it, or chemtrails to you and me. Here in Thanet on the tip of Kent, they started spraying the skies at 4am today. The whole sky is smothered with a white smog and it makes my throat scratchy when I go outside. God only knows what it does to the sea, farmland, animals, rivers and everything it lands upon.

The sheer lunacy of spending billions on solar farms and then reducing the sunlight by 30-40% is beyond a joke.

I would like to know if anyone out there has taken dust samples from cars, fields and buildings to see exactly what does reach the ground and what are the long term implications on food and health and environment? There was no government consultation with the public about this, just a silent agreement of the Paris climate agreement in 1996. The last two years have been awful, no blue skies at all for more than an hour or so.

—*Toby Thompson*

XVI.

Climate engineering's double trouble

By Susan Elaine

> Music and sound have been Susan's passion, growing up and professionally. Enjoying life as a healthy, active and sound sleep person, she found the onslaught of overhead spraying to block the sun was affecting her mind and body. That lead to her researching what was falling on our heads, what we were breathing in, and why it is taking place.

You and I are nature, just like our environment. We need clean air, water and food to survive. It's been that way for centuries.

We entered the third industrial revolution after World War II, when pollution really became a by-product of progress. Issues were developing, but most in industry did not take responsibility for their toxic waste or seriousness of this situation.

Companies are hiding factual data for decades because they shamelessly don't want any conflicts with their marketing or profits. Toxic chemicals, steel production, transportation, coal mining and military are the top five industries polluting natural resources and our health. Let their lawyers do the talking, downplaying involvement and finding loopholes to wiggle through.

The media conversation during the past two decades shifts the blame for transportation pollution and toxic chemical waste by redirecting our attention to global warming.

Global warming boomerangs around the world using weather modification and HAARP, High-frequency Active Auroral Research Program. A hand me down from the US Air Force to University of Alaska, HAARP can change wind patterns and velocity to manipulate weather systems.

We are now being told that extreme weather and heat are the real issues. Climate Engineering becomes the new catch phrase and blocking the sun is the knucklehead idea.

Most don't realize that Climate Engineering adds more pollution to an already high alert situation. Heavy metals, aluminium, barium, strontium, mercury, sulphates, fibres, nano particles and much more, landing on your head and skin. We breathe these toxins into our lungs and eat food crops covered in it.

While distracting us with fear of food shortages and erratic weather conditions, those working behind the scenes are staging these events. Clearly visible but misrepresented with denials until caught. How many lies have diverted our common sense and censored our critical thinking?

Let's check out the clues:

FIRST CLUE: WHY IS NATURE BEING SUFFOCATED, STARVED AND DYING?

Industrial pollution also includes dumping chemicals and toxic waste from manufacturing facilities into air, soil, streams, lakes, rivers and oceans. Factory farming, feeding animals gene edited GMO crops sprayed with glyphosate; housed in horrendous conditions and contributing greatly to greenhouse gas emissions.

Pesticides are poisoning our ecosystems and making the food we eat high risk. Glyphosate causes cancer and is deadly. Our food crops are drenched in these toxins to kill bugs, which also kill bees, worms, microorganisms, frogs and birds needed for the web of life to survive.

Corporations destroying rainforest, cutting down trees and wildfires burning trees from the inside out. Cruise ships dumping their waste. How many cities are polluted with PCB's in their water supply because Monsanto said it was safe. The whole ocean is under siege from toxic chemicals and plastic. Toxins are choking marine life and blocking the natural process of photosynthesis, when healthy plant organisms absorb carbon dioxide.

Did you know there is an island of plastic in the ocean which is twice the size of Texas or three times the size of France? Water currents have pulled this together between California and Hawaii. Those profiting from using plastic containers ignore their product life cycle and have so far refused to take responsibility.

Second clue: What did I observe?

I started noticing overhead spraying in 2014. Friends suggested that I look up at the sky everyday and see what was going on, in plain sight. Back then it was sporadic, but when I did see planes drawing lines in sky, by end of day I would feel flu-like symptoms. Metallic taste was a give away, my nose would burn and run, my eyes burn, scratchy throat and cough were undeniable.

It takes a bit of time to understand what is going on and put it in perspective, but by December 2019, overhead spraying assaults were a daily occurrence. These manoeuvres were ramped up to being unbearable. The air quality was so bad that I had to keep windows closed and an air filter on constantly. I was not going outside unless necessary, always wearing a cap or hat. It became obvious that making people feel sick on purpose was the agenda.

I started noticing how irritated my skin was, waking up

in the middle of the night scratching my head or leg. I soon realized that I could not touch my skin because rubbing made it more irritated. My breathing was heavy from toxins filling my lungs. I would come home from riding my bike for a couple hours on a partly blue sky day, feeling drugged and energy zapped.

I've always been healthy. Eat organic, do cardio, sound-sleep and hydrate, so flu-like symptoms are rare for me. As a vital, sun loving individual living in South California, I was enjoying my pursuit of happiness until that was derailed by the sellers of overhyped flu season, aka Covid-19.

All of a sudden you feel itchy. Do *not* scratch or rub your skin. Friction from your fingers rubbing particles can spread rash on your skin. Safeguard yourself. Use cloth or tissue when touching your face and shower immediately when you arrive home. Brush out your hair and have fresh clothes ready.

Third clue: Why block the sun?

To block the sun is the dumbest idea yet. We need sunlight for vitamin D which helps heal our skin, processing at cellular level. Sunlight is a disinfectant, protective against many viruses and bacteria. Plants and food crops use photosynthesis that need sunlight to accomplish. Farm soil needs sun, water and microbes to regenerate nutrients. Blocking the sun disrupts plant growing cycles, which harms our food supply and farmers' survival.

When overcast or milky cloud cover deflects sunlight, it gets stuck in the stratosphere until it bounces back to the earth on a cloudless afternoon. Heat intensity builds up, compounding deflected sunlight heat. For example, it's 73 degrees Fahrenheit but heat index is very high at 10. Unnatural and burns. The cover-up excuse is that ozone air pollution can cause respiratory health problems including trouble breathing, asthma attacks and lung damage. The reality is, these side effects have been caused by spraying toxins overhead, under guise of Climate Engineering.

FOURTH CLUE: WHY ALL THE DIFFERENT NAMES?

Hiding a true agenda starts with concealing or manipulating data. Using a name that hides your true intention or changing it to confuse people can delay getting caught.

US military called them chemtrails, when invading Vietnam during the sixties. Agent Orange was sprayed overhead to give the sky the appearance of overcast, milky clouds. Fighter planes could then fly above clouds without being noticed. Food crops were destroyed and people living there made sick with respiratory issues and horrendous skin conditions.

Name change occurs when lies and agendas are exposed. Thus, the Harvard Solar Geoengineering Research Program led by David Keith is funded by the same foundation which funds gene edited vaccines and GMO mosquitoes.

Call it what you like, geoengineering, weather modification or manipulation, cloud seeding, lines in the sky, aerosol clouds, chemtrails, weaponized weather or solar engineering describes a cloudy overcast facade of Climate Engineering.

WILL COMMON SENSE AND CRITICAL THINKING RISE ABOVE THIS ORCHESTRATED EVENT?

A better way to focus on climate starts with our food supply. Farming with organic quality and regenerative principles has huge benefits. The end result would be healthier children, with no deadly pesticides contaminating our soil, air, water and food supply. Industry past and present made to clean up the toxic mess and remove plastic from ocean.

Under the guise of Climate Engineering, two seemingly disconnected events were activated in one action of overhead spraying. Blocking the sun by deflecting UV rays with toxic particles which also dispenses flu-like symptoms. You are being make to feel sick, on purpose. My eyes burn and my face hurts. This silent torture is devastating our mental, physical and emotional well-being while destroying our natural environment. Double trouble, takes a bad situation and makes it worse.

Learning that carbon dioxide makes up just 0.04% of the atmosphere was a breakthrough moment.

—*Izzy Solabarrieta*

XVII.

Confessions of a climatard

By Izzy Solabarrieta

> Izzy is a former journalist and TV producer turned small business owner who is proud to match the definition of 'climatard'—defined by *Urban Dictionary* as 'A person who denies climate change is real and happening, in opposition to the vast majority of climate scientists.'

I dislike the word 'climatard', although according to the mainstream media I am one.

I'm a fairly recent convert to this exclusive rank of Official Disbeliever of the man-made climate change narrative.

And it is how I can understand why people find it so very hard to comprehend that there isn't actually a climate emergency going on, at least not as we are told it is.

Unlike my fellow climatard friends, who say it's always been very obvious to them that it isn't happening, and that the official lines make no sense, to me it was the opposite.

Indeed, the whole thing seemed so real—and the arguments made so much sense—that it was nearly impossible to understand how it might not be happening as the mainstream says it is. I still have to occasionally check in with myself, even now, to remind myself of the facts and stay on track —although these days just looking up to the sky is enough.

As did I eventually get there, I hope my experience might help others to understand the huge leap required to go from believer to questioner, and possibly to provide a couple of steps for others along the way.

Life BC—Before climatardism

In my previous existence as a non-climatard—i.e. someone who believed the media's narrative—I've been more active than most.

The climate and its impending disastrous changes have dominated much of my life. Fear, guilt and sorrow hung over my family and I for years.

I first learned about the greenhouse effect at school in the 1980s, when it all seemed a long way off. Over the next two decades it started to become real, a slow drip to get into the psyche. In 2006 I took the summer off work and volunteered with Friends of the Earth, campaigning for climate legislation.

I contacted the TV soap Emmerdale, inviting volunteers from the cast to appear in a short film. Two of the regular actors agreed; I booked a hotel room and filmed them sleeping in bed, wearing comedy pyjamas, waking up to a loud alarm clock and screaming 'Wake up to climate change!' They then did the rounds of morning television to discuss how everyone else needed to wake up too.

A year or so later the big scare story in the news was that we had just 60 months left before catastrophic climate doom, and I made another film with a group of people standing unmoving in Leeds city centre—frozen—to mirror the ice caps.

I was furious when Channel 4 broadcast The Great Global Warming Swindle, unable to comprehend how they could be so irresponsible. I organised viewings of An Inconvenient Truth to counteract the damage. I dismissed the odd person who suggested an alternative to the narrative as dangerous and mostly not worth talking to.

I shared the starving polar bear image widely. I raved

about Greta. For years I refused to fly. Even I made mistakes though. I can still hear the collective gasp uttered by a group of fellow believers after I made the mistake of telling them I had innocently—ignorantly—had my beloved horse cremated.

Fast forward a decade, several hot summers, and into the dry, scorching heatwave of 2018. Around me everyone was loving the good weather, the drought unnoticed because their televisions weren't telling them there was one, yet my heart was heavy. I wondered how people could not notice what was happening, or care so little about it.

My heart broke every time I walked past our local nature-filled pond, which became a dried-up stink hole from early July onwards. I doubted I would see it full ever again—I have done, of course, every year since.

I began to realise sadly that we humans, with our selfish, frivolous living, could have no future on this planet without being seriously controlled. And yet I loved life, loved to eat nice things, buy clothes, drive my car, go places. The inner debate made me miserable.

GENTLE AWAKENING

Fortunately, sometime around 2012 over after-work drinks, a colleague had casually mentioned a conspiracy over the New York 9/11 terror attacks. It was my first foray into alternative thinking, but it sparked a gentle awareness which grew over the coming years. I evolved from believing the nice people on the TV, to being reasonably awake—and stopping watching television completely, which I would argue is the best thing anyone can do for themselves.

But even through 2020, the climate nut refused to crack, despite knowing about the globalist agenda for control. The world seemed to be obviously getting hotter, it didn't make sense for it not to be true. And humans use so much oil, it seemed impossible that we were not doing any damage.

I was blessed that my climatard friends—climate deniers! —would occasionally post, and I did read their posts and memes. I found myself reading through and suddenly agreeing with the lists of climate alarms which had not come true.

Learning that carbon dioxide makes up just 0.04% of the atmosphere was a breakthrough moment. So was watching the clip of Dan Pena in which he asked why the banks were still doing mortgages for those areas which would soon be under water. And also why billionaires were still buying beachside mansions.

Photos side by side of 'before and after' climate change— with, very obviously, zero sea level rise, were also powerful.

Being reminded that CO_2 is plant food helped. I had learned this in school, but it made sense that our over-use of 'fossil fuels' had knocked the balance.

And then I read a few twitter conversations suggesting that fossil fuels might not be quite what we've been told they are. I learned how supposedly empty oil wells refill, but no one in the industry is allowed to talk about it.

There were tweets stating that oil is nothing more than vegetable matter. Others claimed we have been lied to about dinosaurs, whose bones are supposedly the source of oil. I still don't know what is real, but it certainly all got me thinking.

And then there was God. I U-turned back to church in 2020 as it was clear this situation was too big for humans. As I read the words of Paul, Acts 17:28, explaining how we live, move and have our being in God, I found myself wondering how it was even possible that there could be imbalance and doom in God's Kingdom.

A year or two ago, finally watching the Great Global Warming Swindle was definitely better late than never! At last, things made sense. I finally understood what was really happening and what the science was to destroy the hoax. Now that I understood, I could share it with others.

I've recently also watched Climate: The Movie. Both films are packed with great information and I have made notes to help me in my discussions. I also use it to remind myself—it turns out it's a lonely world being a 'climate denier'. 'Climatards' have lots more friends, which is fine by me. I have no interest in being around group think.

LIFE AS A CLIMATARD

Since my full awakening I have been pretty vocal about it all being a hoax. Barely anyone challenges me, sometimes I think that's because they think I am crazy, but mostly they probably don't have any facts to come back at me.

And this just confirms absolutely, disappointingly, that only a tiny percentage of people will ever do any research, into anything!

Yet I find that, for us, being armed with knowledge is absolutely key. If we don't know what we are talking about, the hypnotised TV watchers might not have facts but will immediately spot our weakness.

I have found one potential way in is to say that I cannot have a conversation about climate change while no one is talking about the mass plane trails in the sky, and what effect that is having on the climate and weather. I have found that some people have noticed them, perhaps without realising it.

Recently, an American woman challenged me, saying she had seen the disappearing glaciers in Montana, clearly evidence of climate change. When I said I had seen terrible footage of heavy, constant plane trails over that state she huffed up. I don't expect her to change her thinking from our conversation, but surely something must happen in your head when you know the other person has said something legitimate. No matter how hypnotised you are. Hypnosis can be broken.

There are some positive signs.

I decided to stand in the latest general election for the Heritage Party, which is taking a stand against Net Zero and the climate nonsense. I received some unpleasant comments and emails, but the vast majority of feedback was good. After my leaflet, which called for a climate debate and questioned Net Zero, was distributed around the constituency, numerous people quietly came up to me and said they appreciated what I was standing for.

I recently challenged a believer on Facebook, who commented about the world being supposedly hotter than it ever has been. My reply, saying 'sorry but that is absolute nonsense, the world has been much hotter than it is now, many times' got an astounding five likes! I don't think that would have happened even a year ago. And while those people might not be brave or confident enough to comment, they have seen that I did and read what I said.

Ultimately, we have to know that good—Truth—will win. Truth has already won. Humanity can and will awaken. Because, after all, to quote John Knox, one with God is a majority.

I have just received plans from Stockport Council and have been watching this take place before my eyes on how it's becoming a 15-minute city.

Plans are to build thousands of homes in disused mills around the town centre which right now are in a fast state of construction, which they say will be a fully walkable neighbourhood that's connected to the town centre and surrounding communities. These buildings will have on site shops, dentists, pharmacies, gyms, meeting rooms, cafés. In close by areas there will be dedicated walking and cycle lanes and pedestrian streets only. Their plan is to be completely carbon free by 2038.

If this is not aimed at being a 15-minute city then what is? And the townspeople are walking into it blindly, believing it's purely for their benefit.

All this of course is already surrounded by hundreds of cameras watching your every move all over town.

—James McGuinn

XVIII.

Defeating the narrative of fear

By Cherry

> Cherry trained as a nurse 37 years ago and has worked at an
> advanced level for over 15 years. She bore witness during Cov-
> id-19 and saw the devastation to people's health caused by the
> lies and subsequent faux 'treatments'. She loves her family—she
> has three children and two grandchildren and that's why she
> speaks up and will continue to her last breath.

We face a crisis, not from 'climate change' but of the reaction to it. We need to enter the Hegelian dialectic
here, so let's go where many fear to tread...

So called 'man-made' climate change, is referred to as 'an
existential threat to humanity', but it is not. Many governments have been captured and through their successive creeping legislative changes, this capture has been passed down the
chain to the local level.

Our world has changed and not in a good way. Societies
around the globe are hamstrung by rules and regulations, not
least by the 17 so-called 'sustainable development goals', that
are anything but, and packaged, as they are, in bright colours
with friendly graphics, presumably to appeal to the younger,
and they feel, 'more malleable' in society. One could postulate
that these goals look like a plausible way to 'save us all from
ourselves'. Sadly, this could not be farther from the truth. The
reality, for anyone who has eyes to see, is that these stated

goals spell the end for humanity as we now perceive it, save for the elite, who are protected and sheltered from the most draconian set of objectives ever put together by a group of sadists, hell-bent on their vision of 'the new world order' they so desperately crave. By following these ludicrous targets we condemn the majority of the planetary population and its animals to a frugal, hollow, desperate future, devoid of love, happiness and hope or anything approaching what we may term as 'free will' or 'individual choice'.

The world of our youth has changed and will continue to deteriorate, not because of a change in temperature but by the actions of the deluded.

What are we told is the driver of the imagined destruction of our planet? The answer is shouted from the rooftops ... 'Carbon dioxide!'

Think about any time that you spend outdoors and you will see growth, life and beauty. Trees that reach up majestically into the air, branches swathed in greenery and flowers, tree canopies nested by birds and squirrels scurrying from one branch to another. One could marvel at the sheer awe and beauty of it all.

So what is the fundamental element that is required to achieve this abundance of life in all of its majesty and beauty?

Well ... that would be, carbon dioxide.

Those of us of a certain age can all remember our early science classes. They taught us the mechanics of photosynthesis being the driver for the life cycle of any living plant. This was information fundamental to understanding our ecosystem at the most basic level. The process by which we grow crops, care for the soil, feed ourselves, ensure good health and nurture our children. We all understand this, right? Then how do we explain the Orwellian push to banish carbon like some errant child who has misbehaved in class? Could it be that we are forgetting that, we ourselves are carbon. Maybe we should banish ourselves? Oh, I see, this is actually what they want.

Now it's all falling into place...

There are some almighty untruths circulating about this whole 'global climate boiling change saga' and those lies have driven into town in souped-up monster trucks, knocked down the walls of the house, dragged us screaming from our beds and proceeded to obliterate any documentation, study, or real world evidence that may dare to set the record straight. This is done in the fear that the truth will out and the whole scam and intricately woven tissue of lies will scatter to the winds. That cannot be allowed to happen.

This lie is far bigger and more brazen than 'Covid' could ever seek to be. 'Covid' was the preamble, the warm-up act, the entrée. The CO_2 lie is used to depopulate the planet, and the possibilities are endless. They include (and this is not an exhaustive list): control of where we go, what we eat, how and where we travel, communications, how we occupy our leisure time, car ownership, owning pets and livestock, the supply chain, farming, ecosystems, employment, procreation, and so on and so forth.

This lie has been crafted, dictated, managed and implemented at the global level, by such insidious organisations as the WHO, WEF and multiple NGO's too numerous to name here, however, you can be sure that the boy Gates has interests in any number of these. The lie is packaged up and fully adopted by almost all governments of the world, then, in order to embed and spread its tentacles through every aspect of our day-to-day life and work, it is discussed in local council meetings, adopted and marshalled onward by numerous 'little folk', buoyed up and drunk on any degree of power in the form of a fancy title or perhaps a mayoral chain. These people wield the authority to dictate to the little people (us), that ... 'they had better adhere to the latest, unscientific and nonsensical pronouncement, or the blighters will create such chaos that we will all die, the sky will fall in and the planet will set ablaze and burn for all eternity, The End'.

The fabrication that carbon is the protagonist that needs to be removed from our lives plays out in the following malign plans; Net Zero, 15-minute cities, C40 cities, carbon capture, rewilding, solar and wind energy, electric vehicles, draconian restrictions to farming, food production and transport infrastructure to name but a few.

It seems that the bigger the lie, the more difficult it is to refute. There are certain lies that have been taken out of context and used to prove the false case for man-made climate change, even when they are easily provably false. As an example, the University of East Anglia and its 'Climategate' scandal, where it was clearly demonstrated that scientific data (that did not fit the narrative) had been manipulated and changed to present the only conclusion that is allowed.

A quick search on the internet today meant that I was reliably informed by an AI generated answer, that this event has been 'fact checked' so we can be reassured that no data manipulation ever took place...

The other commonly reported statement that '99% of all scientists agree that carbon increase is the cause of man-made climate change' is another lie. This is reported ad nauseam by the media, politicians and others in order to cement in the public consciousness that this is an accepted and unarguable truth, when this too is an intricate and deliberate distortion of the true facts, so manipulated yet so easy to get people to buy into.

The response from the normie population? 'They wouldn't announce that if it wasn't true!'

It is really difficult when you know that this is such a massive lie and manipulation, to convince those who receive all of their 'facts' from the tel-LIE-vision that they have been duped.

There is a wealth of evidence for example from extensive study of ice core samples that clearly and indisputably show that carbon dioxide increases follow warming (after a delay of circa 800 years), *not* the other way around. Carbon dioxide does *not* drive warming. Temperature records that are paraded to the public to make the case for the evils of this demon gas, fail to show that we are in a steady climb out of the last big cooling period, known as the 'little ice age' and that the earth is now in one of the coolest periods in its history. A history that shows very clearly the decreases and increases in temperature, far greater that we are experiencing today, that were quite obviously not driven by us pesky humans and our industry, on the contrary, we have survived and thrived despite those enormous swings in temperature. The rhetoric that 'one degree of warming' is catastrophic and that whole areas will be flooded and under water because of this, are laughable. The notion that melting ice caps will cause the sea level to rise, is equally nonsensical because the weight of the ice in the water is the same as the volume of the melted water, they are equal. This is basic science.

If we allow these maniacs to continue unchecked we will suffer, and it will not be because of a rise in temperature, it will be because sanity has been replaced by insanity and totalitarian rule has replaced free thought and individual freedom. This is what is known as slavery, pure and simple. We have no choice but to say no to this and resist at every opportunity all the changes that drive us toward this dystopia. Do not fall for the mantra that if you care about the planet you will recycle your rubbish, when they wrap everything in plastic at the supermarket, when they provide milk and juice in plastic cartons instead of environment friendly options like glass, which can be cleaned and reused; this is what we used to have until they changed it. We have used increasingly economical and clean diesel cars, but they said we need to pillage the third world for lithium, we need to use child and slave labour to fulfil the EV dream and ship them and their components all over the world in huge tankers. We increasingly see potholes and damaged roads because of the weight of these EV monsters. We ruin prime farming land with solar panels and wind turbines which blight the countryside, destroy wildlife, reduce food production and push up food prices, inevitably increasing the risk of a global famine. We dig into the seabed and pour metric tonnes of concrete to prop up these superstructures, which, when their short life span is over, have to be blasted out of the ground and buried, because they are non-biodegradable and are unable to be reused. Huge graveyards are dug to bury these monstrosities, further degrading the land and who knows what leaches out of them into the soil and the water table.

Fossil fuels are not dangerous to us, they are compressed foliage, they are already made, they are available, they are cheap, they do not contribute to climate change. Oil is *not* a fossil fuel, despite what you may have been led to believe, it exists below the level of fossilisation and is widely acknowledged to be in constant production and abundance, it is *not* a scarce resource. As individual sovereign beings you need to look into this yourself; 'do your own research' is not a dirty word. Do not believe what you are told just because it is mainstream, you should be sceptical because it is mainstream.

In summary, these people, the perpetrators, are mad, these policies are mad, and if you have any thought that this is the way to a 'greener, more sustainable future' for this planet, then I have to refer you to the words of Mark Twain who said: 'It is easier to fool people, than to convince them that they have been fooled'.

The manifestation of wildfires is reported by MSM in such a way as to convey the illusory conclusion that they are caused by global warming. That is blatantly untrue. Most fires are caused by human activity either deliberately or accidentally.

—Louis Brothnias

XIX.

Anthropogenic climate change— scientific fraud, part two

By Louis Brothnias

The very long-term (slow) Milankovitch cycles describe the ever-changing Earth motion/ position as the planet orbits the Sun, but do not account for the alleged (claimed but unverified) rapid escalation in global temperatures. How are these alleged escalating temperatures measured? The positioning of instruments to monitor the temperature at a specific location is critical. Is the monitor placed at ground level in an ever-expanding concrete jungle (city location)? The more concrete, the greater the heat capture. Or at an airport that has high volumes of hot aeroplane exhaust gases in the surrounding local atmosphere? An instrument placed at an elevated level above the sea or open country (weather balloon) to monitor the upper atmosphere will collect very different data. Groundwater extracted for irrigation and other human activities has displaced 2.15 trillion tons (2,150 gigatons) of water between 1993 and 2010 and has had a significant impact on Earth's tilt (obliquity). Only a very small shift (to <23.5 °) in the obliquity would be enough to result in more land directly facing the Sun and for the quantity of solar radiation captured to increase. The upward movement of radiated heat would then warm the atmosphere. Canada and North America, North Africa, Europe, Russia, China, and India form the most significant land masses in the northern hemisphere, with South Africa, Australia, and South America in the southern hemisphere. Highly speculative statements about the imminent tipping point that would result in the collapse of Earth ocean systems is scaremongering.

The 'agreed' genocidal policy (a protocol) of Net Zero is a lethal plan (a deliberate action) based on a lie. That any further increase in the CO_2 concentration in the atmosphere will lead to a climate crisis is an absurd notion. Without carbon dioxide, all life on Earth would cease. Living entities (trees and plants) convert atmospheric CO_2 into new O_2 in equal amounts (photosynthesis).

The manifestation of wildfires is reported by MSM in such a way as to convey the illusory conclusion that they are caused by global warming. That is blatantly untrue. Most fires are caused by human activity either deliberately (arson) or accidentally (carelessness). A fire once started can be sustained by its environment (marshland, wetland, peat-rich soil = methane). The 1925–1926 wildfire season caused terrible destruction. But almost 100 years later MSM attempts to promote the case for a climate crisis. Bill Gates is one of the wealthiest individuals on Earth (money talks) and pontificates (the self-appointed climate change guru who has no expertise—he's a college dropout) about how to avoid a climate disaster. What is not mentioned is that Gates is the biggest private owner of farmland in the United States. Gates advocates laboratory grown meat and could interfere with the global production of livestock by buying up even more and more farmland to prevent anyone from raising cattle. And in doing so, in the false race to Net Zero, will force people to survive by consuming the fake laboratory manufactured (synthetic) meat which he will supply and control. The scope for genetic adulteration of the meat is unlimited. One man feeding the World and saving humanity. The Messiah complex.

And making $billions and $trillions in the process.

The urge to save humanity is almost always a false face for the urge to rule it.—H. L. Mencken

Electricity has become an absolute dependency—rechargeable devices, electric vehicles (EVs), heat pumps. The natural resource that contains the lithium, nickel and cobalt to produce EV rechargeable batteries is already relatively scarce (2024) compared to confirmed crude oil reserves. The length of time that EVs could exist is extremely limited. Just a few years at most and it's blindingly obvious. Enormous amounts of mined ore are needed to process enough lithium, nickel and cobalt to manufacture just a single average-sized 500kg car battery. The use of oil-based fuels would be necessary to facilitate the extraction and isolation processes. A 10–12 gallon (US/imperial) car fuel tank would hold around 50kg petrol. Crude oil has myriad uses and the petrol collected from the volatile distillates of the refining process is a waste product. The huge battery weight defines a large vehicle to structurally house such a monstrosity. Around half-a-tonne dead weight is necessary to provide the energy to make an EV function. A heavy vehicle has a high inertia and considerable energy is necessary each time it is moved from standstill and the subsequent acceleration to operating speed. The frictional wear on tyres will be massive and they will need to be replaced more often than for the much lighter petrol or diesel-fuelled equivalent. Micro-particles of rubber in the atmosphere will create severe pollution issues, as will the brakes that will wear out very quickly too and create even more atmospheric pollution during the EV's serviceable life. And what happens in the event of a power outage? How can an EV be energised? Solar and wind power alone could never supply the necessary energy—imagine a dull and windless day in Winter. Neither has there been any new-build nuclear power stations. The use of a wind turbine may produce so-called green energy (no CO_2) but the manufacture of the steel to build it does. This applies also to EVs. The major supplier of wind turbines is China and the Chinese intend to continue building coal-fired power stations and the reason should be glaringly obvious. Wind turbines and EVs may be regarded as environmentally friendly in use (do not produce CO_2) but the manufacture of both produces an enormous yield of CO_2.

Heat pumps are claimed to extract warm air from an outdoor atmosphere and transfer it indoors, though this requires electricity to function. Heat is not actually generated. Just extracted and transferred. On very cold days conventional oil-based fuels would be necessary. And no doubt some authority would declare when it's cold enough and give permission to use conventional energy sources. Or so one would expect.

Venus is very, very hot. Scaremongering (mainstream media) persists by claiming that Earth is in imminent danger of becoming another Venus. This planet is 40 million km closer to the Sun than Earth and has an atmosphere of 95% (950,000ppm) CO_2. The concentration of CO_2 in the Venusian atmosphere is $95/0.0412 = 2300$ times that of Earth. The natural difference in temperature between Earth and Venus is enormous (464 °C – 15 °C = 449 °C) and is far in excess of the temperature to melt lead (327.5 °C). Sources other than atmospheric carbon dioxide for alleged 'global warming' (on Earth) must be sought elsewhere.

Carbon dioxide is not the cause of alleged terrestrial global warming.

The doomsters' favourite subject today is climate change. This has a number of attractions for them. First, the science is extremely obscure so they cannot easily be proved wrong. Second, we all have ideas about the weather: traditionally, the English on first acquaintance talk of little else. Third, since clearly no plan to alter climate could be considered on anything but a global scale, it provides a marvellous excuse for worldwide, supra-national socialism. All this suggests a degree of calculation. Yet perhaps that is to miss half the point. Rather, as it was said of Hamlet that there was method in his madness, so one feels that in the case of some of the gloomier alarmists there is a large amount of madness in their method.

The fact that seasoned politicians can say such ridiculous things [about reordering civilization]—and get away with it —illustrates the degree to which the new dogma about climate change has swept through the left-of-centre governing classes...

—Margaret Thatcher, in her autobiography 'Statecraft: Strategies for a Changing World', 2002, pp 449–50

XX.

Dreamer

By Ray Wilson

> Travelling around the world with his family as a young boy, Ray became fascinated by radio technology and electronics. This led him to pursue a career in engineering, where he tried to make a difference and in small ways challenge the regime that controls us all. When he's not working with electronics or writing stuff, you'll probably find him out motorcycling with his wife, walking the dog, or working on various projects. He finds solace in the simplicity of nature and the freedom of the open road.

Peanut butter logic, served on a bed of lies.
Don't go down too easy when you've seen your father cry.
Have you ever clutched the steering wheel in your car too tight?
Praying that police sirens just pass you by that night.
While the helicopter circles us, this theory's getting deep.
Think they're spraying chemicals over the city.
While we sleep.

Come on, I'm staying awake.
You can call me a dreamer too.
(I got one eye open for these devils.)

—Prince

Sometime earlier in the 1970s, in a dimly lit corner of their private club, two of a small cluster of diabolical figures leaned in close.

'We need a new enemy to unite us,' the first figure said, a sly smile creeping across his face. 'In our search, we hit upon things such as pollution, global warming, water shortages, famine ... all these could serve our purpose.'

The second figure nodded thoughtfully, his eyes narrowing. 'Indeed, in their totality and interactions, they form a formidable foe. These issues are universal and demand the solidarity of all peoples.'

'Exactly,' the first figure continued. 'But by designating these as the enemy, we risk falling into a trap we've warned ourselves about before. We could be mistaking symptoms for causes.'

The second figure raised an eyebrow. 'Go on.'

'All these dangers,' the first figure explained, 'are results of human actions. It's only through changing attitudes and behaviours that we can overcome them. The real enemy, then, is humanity itself, and that is what we will make them believe.'

A sinister silence fell between them, each contemplating the weight of this revelation. The second figure finally spoke, a dark gleam in his eyes. 'If humanity is the real enemy, then our strategy must focus on altering human behaviour. But how?'

The first figure leaned back, a devilish grin spreading across his face. 'By manipulating fears and directing efforts towards these symptoms, we can control the narrative. Make them believe they are fighting pollution, global warming, and the like, while we subtly guide their actions towards the changes we desire using clandestine weather manipulation.'

The second figure smirked, raising his glass. 'To a new enemy, and to our new plan.'

'To humanity and culling of the useless eaters,' the first figure echoed, clinking his glass against the other. 'The true adversary is hidden in plain sight.'

The city of 2050 bristles with biometric surveillance eyes and incapacitating LED street lights with infra-sonic sound weapons.

'Googles and defenders now!' someone shouts.

Half a dozen lithe figures clad in dark Faraday suits shimmed up the poles with lazer cutters poised.

The first poles were severed; the other poles activated multiple incapacitators shooting out pulsed beams of light in defence. The infrasonic sound weapons deployed eardrum-bursting levels of energy before they too fell silent. The figures swiftly disabled the surveillance devices and LED street lights, plunging the city into darkness. As they disappeared into the shadows, drones appeared above the buildings, locking onto their DNA targets.

'Drone Coders, let's go,' a voice shouted. With their mission accomplished, the figures vanished into the night, leaving no trace behind.

Lara's heart pounded as she navigated her way through the labyrinthine alleys of the city. The towering digital billboards above blared the same ominous message: 'Stay home. Stay safe. No, Planet B. Stay in to help out.' Images of desolate landscapes, scorched earth, and dying animals flashed across the screens: animals clinging to branches, scenes of flash floods—buildings swept away, uprooted trees torn from the earth by tornadoes—the apocalypse that gripped humanity.

Lara slipped silently across the wasteland towards the tenement building.

'Hi there, Joe, What are you doing up so early?'

Joe sucked on his cigarette, blowing rings of smoke. 'I couldn't sleep,' he replied, his voice hoarse from the early hour. 'I like to watch the skies; no one looks up these days; they are always looking down.'

'I've seen quite a bit of change in my time,' old Joe said, looking back at Lara with a sense of nostalgia.

Lara climbed over the balcony rails.

'There are a lot of climosnitchers and climopreneurs around here.' Lara pointed to the apartment block across a cracked and pitted asphalt road.

'I've heard a bit about those climopreneurs making quick bucks out of carbon credits, but I have no time for the climosnitching-curtain twitchers we used to call them. I prefer to keep to myself and focus on my own business,' he said, taking another drag of his cigarette. 'But it's hard to avoid all the gossip around here.' Lara nodded in agreement, gazing out at the changing skyline as the sun began to rise.

'The climate marshals are already out on patrol.' Lara pointed to a couple of figures strutting along the road, looking for those not adhering to climate-friendly practices.

'Their sniffers will detect your smoke, won't they?'

'They're out to cause trouble,' Joe replied. 'You want some water, Lara? I got my alkaline ioniser and water filtration unit working.' Joe poured out a cupful. 'I can't drink the bowser stuff; it's full of chemicals; it makes me sick even if I distil it.'

'Thanks, Joe,' Lara said, taking a sip of the filtered water.

'I've got to keep it quiet,' Joe chuckled. 'I can't trust anything these days.'

'Back in the old days, they were talking about the chemtrail flu; everyone in the neighbourhood started fighting after the aeroplanes flew over; lots got sick and started puking up.'

'So you still like to watch the skies, Joe?'

'I watch—I see the polymer gloop and graphene oxide. I got my dosimeter right here,' he said, pulling the detector out of his pocket. 'I have cumulative levels; I have to keep raising the ambient alarm levels; otherwise, it's beeping all day long.'

'What was it like, Joe? Were there lots of doomerists back then?'

'Not so much—there were grisly acts, suicides, and fearless acts by those convinced that the world was about to end —they felt compelled to end their lives and thereby reduce the surplus carbon.

'Not overt disaster capitalism—there weren't euthanasia exit hubs back then,' Joe explains.

'Ah, I see.' Lara stretched out her legs.

'My mum refused to have the vaccine in the 2020s, but they got her in the end with the aerosolized gene therapy and put her on the authorities database.

'They think they own us all, but they don't,' Joe said, lighting up another cigarette. 'Just because they slipped luciferase into some folk's DNA means diddly-squat.'

'Did you get the augmented gene, Lara?' Joe looked away and said, 'Sorry, I shouldn't have asked—it's my old peanut butter logic and it's spread thin.'

'See you later, Joe.' Lara waved goodbye as she hopped over the balcony and walked away, leaving Joe alone with his thoughts. As she disappeared around the corner, Joe took a long drag of his cigarette and exhaled slowly, contemplating the world they lived in.

The diabolicals believed that after years of predictive programming, they had convinced the masses that their actions had brought about this cataclysm. The once vibrant society was now paralysed with fear and guilt and dumbed down with chemicals. The mandated emergency vaccines supposed to 'mitigate' so-called climate sickness just made things worse; ironically, they caused a surge in autism among the population, a side effect conveniently ignored by the government and the media.

Lara knew better. She had seen the unaltered data, the hidden reports, and the truths buried beneath layers of deception. She knew the planet wasn't boiling because of the common people's actions; it was the deliberate manipulation of the environment by the diabolicals, the elite, who thrived on chaos and control.

In her small, dimly lit tenement apartment, Lara had amassed a treasure trove of forbidden knowledge. Holographic projections of suppressed research papers and censored news articles floated around her. Each piece of information was a weapon, a tool to fight back against the propaganda machine that had ensnared the world.

Her most valuable possession was a crude, makeshift CB transmitter cobbled together from discarded tech and salvaged parts. It was her lifeline to the underground network of truth-seekers, the only connection to others who resisted the diabolical iron grip.

One evening, as the city's curfew alarm sounded, Lara received an encoded message through her transceiver. It was from Max, a fellow resistance member and a brilliant hacker. Max understood blockchain technology—he had found its Achilles heel—one change somewhere is a change everywhere. His message was terse but loaded with urgency: 'Meet at the old library. Midnight. We have a plan.'

Lara slipped into the shadows, her heart heavy with anticipation. The old library was a relic of the past, a place where knowledge was once freely accessible; now, compression secure steel screens excluded curious minds.

As she approached the dilapidated library, Lara spotted Max and a few others huddled together. Max's eyes gleamed with a mix of determination and fear. 'We've found a way to broadcast the truth,' he said, holding up a small device. 'This can override the main broadcast signal, but we need to get it to the central tower.'

The central tower was the heart of the diabolical propaganda machine, heavily guarded and nearly impenetrable.

The plan was audacious, bordering on suicidal, but it was their only hope.

With meticulous precision, they outlined their strategy. Each member had a role to play, and failure was not an option. As the clock struck midnight, they set their plan into motion.

Lara's pulse quickened as she navigated the darkened streets, evading patrols and surveillance drones. She felt a surge of adrenaline and fear, but also a fierce resolve. The world needed to know the truth, and she was willing to risk everything for it.

Reaching the tower, the group split up, each taking a different route to their designated positions. Lara's task was to disable the security system, a feat that required both her technical prowess and nerves of steel.

Inside the tower, the air was thick with tension. Lara moved swiftly, her fingers dancing across the control panels. With a final keystroke, the security system went offline, and the tower was vulnerable, but only for a minute.

The others moved in, planting the device at the core of the broadcast system. As they activated it, the screens across the city flickered, then displayed a new message: 'You've been lied to. You are not the problem; every one of you is the solution. The world is not boiling. The diabolicals are the true enemy; you are the carbon they want to get rid of.'

Linked to the blockchain, every screen, billboard, and holographic generator in every smart city and ten-minute town around the world broadcasted the same message at the same time in the language of the region.

On the other side of the city, Joe sat on his balcony, smoking cigarettes and flicking ash, looking into the night sky. He watched as people ran from their tenement blocks into the streets. Men and women, boys and girls, stood looking expectant, staring upward into the sky. Some were scared, some confused, but all were searching for answers. Joe knew that chaos was about to erupt in the city, and he had a feeling that things would never be the same again. Joe took another drag of his cigarette, contemplating the implications of this revelation for the world he thought he knew.

Our species has already lived through both warming periods and ice ages and managed to survive and thrive again even without the modern technology we have today.

—Eileen Coyne

XXI.

Feel the heartbeat

By Eileen Coyne

Eileen Coyne first had poetry published with the Letterfrack
Writers Group in the late 1990s and is author of Oubliette.
Eileen lives and works in Spain. She is a human rights activist
and keen environmentalist. She donated land in Salruck, Con-
nemara to Green Sod Ireland, a Land Trust charity, in 2013.

I went for a walk in the park that Spring morning as I was
sick of my own company, sick of the stale air in my flat
and of my own relentless mental circling and constant internet searches. I was looking for a eureka moment from tainted
search engine modulations. My physical strength was as fra-
gile as my mental and emotional state. Even the simple act of
walking seemed an effort, my feet dragging and threatening
to trip me. I went for a walk in the park that morning and
found a friend.

It had not been my intention to talk to anyone. Like a
true Londoner, I left my flat armed with a book so that, if the
sunshine lasted, I could sit on a park bench and read, or pretend to read, thus not running the risk of making eye contact
with anyone. I found a good spot, near the pond, reading very
little but listening to the birds, smelling the freshly mown
grass and feeling the promise of summer from the soft kisses
of sunshine on my face.

I was feeling close to contentment when I noticed a pair

of legs standing stationary before me. My defensive hackles rose. A nutter or just a lonely person desperate for a few minutes of interaction? My face automatically moulded itself into hard lines, my eyes steely as I looked up to examine the man standing over me. Would his eyes be full of gloating evil, or wildly wandering or pleading? My swift upward glance had already confirmed that, unless he had a weapon, he was unlikely to prove a physical threat. My gaze met with a pair of hazel eyes, filled with calm, compassion and warmth. An older man, in his late sixties, white haired and small of build but he had an air about him; a strength, a confidence, he carried a sense of peace with him. My body relaxed even before my brain had finished its judgement of the situation and, when he explained that he normally sat on that particular bench each morning and would I mind if he sat there also, surprisingly, I said he could and I actually meant it.

And, as the days grew longer and the sun shone stronger, so our meetings on the park bench flowered into friendship. I would turn up two or three days a week, depending on my work deadlines, and he always wandered in around 11.00 am. Sometimes, I would bring coffee for us and he sometimes reciprocated with some tasty morsels from the local bakery. We looked for common ground; I was football and he was cricket so sport was rarely discussed, apart from a shared love of boxing. What we did have in common was a mutual interest in current affairs. I had been anti-vax from the start, he had received one jab and then a neighbour had died soon after receiving one which had made him think and do some research. We discussed possible motives behind the Covid-19 years and none of the possible outcomes seemed like good news to the working man or woman. Indeed, we both agreed that some ugly agenda had been unmasked. The Health Service, not that you could trust it any more anyway, was as good as gone. Globally, absurd measures were being rolled out with a seemingly complete disregard for human rights or respect for the individual. Nothing made sense any more. My friend, with his calm, clear gaze could see that I was a bit at sea, jumping at shadows and looking to identify the next threat before it landed. Whilst I railed against the cost of living crisis, artificial intelligence, threats to cash and CBDCs, global warming, gender modification for children, warmongering with suspiciously high numbers of fighting aged men being flown into various countries for unknown reasons and euthanasia being rolled out in some countries and even being offered to people who were temporarily depressed, he sat calmly on his side of the bench and, being the good friend that he was, he let me vent, throwing an occasional crust of bread to the ducks.

But he was a good speaker as well as a good listener. When he talked, he had the knack of taking you to the place or the person or the situation; you could really visualise it. One day, he started talking about his experience of boxing. He had been an amateur featherweight boxer in his young years. He talked about the first round or two being more to circle and get a feel for your opponent; that everyone had a favourite blow, left or right hand, uppercut, hook, jab or cross; that when an opponent attacked, it was also his time of greatest vulnerability and that, if you could find his pattern, stay alert and light on your feet, you would in all likelihood be the victor, even if he was perceived as having the advantage. There was true artistry to proper boxing. Feints to make an opponent react a certain way. Sometimes quick and savage attacks and sometimes fencing through numerous rounds to wear an opponent down. The importance of proper defence. Suddenly, his soliloquy stopped and I looked sharply at him. He was sitting back, relaxed, with a broad smile on his face. The penny dropped, then, and I truly loved him at that moment. I loved him because, with a complete lack of ego and with kindness, he had unveiled a mechanism that he knew I could follow that would help carry me through the turbulent days ahead. I had complete clarity at that moment. The circling of the first couple of rounds had already occurred in past decades and Covid-19 had been, say, a hard right opening jab or, at least, the first real blow that I had felt. I grinned back at him; he knew it had hit home.

'It is hard' he said 'to keep your footing when every old, familiar assumption has fallen away, but I always repeat this mantra to try and keep me on track:

God, grant me
The serenity to accept the things I cannot change,
The courage to change the things I can,
And the wisdom to know the difference.

Don't waste your energy worrying about what Governments are saying on global warming, for example. It is a feint. When you go home, google 'historical ice ages' and hit 'images' and you will see the great heartbeat of our planet and that there are natural cycles to heating and cooling of the earth every 50,000 years or so. What experts call Modern Man has been on Earth for 300,000 years and that means, my friend, that our species has already lived through both warming periods and ice ages and managed to survive and thrive again even without the modern technology we have today. Centre yourself and feel the heartbeat of our planet; it is a broken world, without doubt, but it is still a thing of beauty and will long outlast all that is worst and best in mankind.'

So what's behind all this fear mongering, 'net zero' and CO_2 propaganda? We know that carbon dioxide is essential to life and plants. CO_2 keeps the planet's temperature stable in order to sustain life. Since it began, planet earth has undergone periods of climate change—some warm, some colder periods but this is mainly the result of solar activity and possible volcanic eruptions plus the oceans which release some $CO2$—an invisible gas which is only 1% of so-called greenhouse gases.

—*Cabochon*

XXII.

Climate change fraud—a grassroots view

By Cabochon

Cabochon has a career in foreign language teaching followed by several years in the health food industry. Further studies and a qualification in human nutrition led to a deeper understanding of the environmental and global forces that shape the health, wealth and wellbeing of the many while enriching the few.

My food waste container/ refuse bins are now rarely emptied on the allotted day. Deep potholes in my street have not been repaired for at least five years. Well-organised local resident protest meetings against demolishing our only local pub/ restaurant failed to halt Council approval of yet another student apartment block in our area. Our city centre, formerly a World Heritage site, is now a world hamburger site, where the homeless shelter in doorways of abandoned department stores, interspersed with the occasional American candy shop to fuel the burgeoning diabetes industry.

Pedestrianised streets of historical interest force public transport into permitted traffic lanes resulting in delays and bus chaos while the war on the motorist sends clear messages via stiff fines, diversions, street closures, expensively created but mainly empty cycle paths and 'low emission' zones. The curse of over-tourism has trashed our city for visitor and resident alike, a fact the local press commentators are aware of: 'Prioritising tourism over the city's own residents—growth for growth's sake does not always serve the needs of the taxpayers.'—*Scotsman*, 30 July 2024. Meanwhile our supposedly 'cash-strapped' local authority complains it cannot afford its traditional duties as financed by council tax while newly elected central government threatens to make up funding deficits by taxing pensions, (some of which were the outcome of salary sacrifice while working) and removing the pensioner's annual heating allowance. Perhaps local authorities should cease expensive theorising about the cosmos to concentrate on local amenities, roads, schools, affordable housing and deprived children in care or special educational needs.

'The aim of present day politics is to keep the populace alarmed and hence clamorous to be led to safety by an endless series of hobgoblins, all of them imaginary' (*Technocracy: the Hard Road to World Order* by Patrick Wood). The bioweapons of modern wars are not artillery and firebombs but fearmongering, misinformation and cognitive dissonance. How else could even highly educated brains accept the impossible as truth such as lockdowns, masking, social distancing and Net Zero nonsense over common sense and reasoning while stifling debate? How else could lamestream media succeed in catastrophising even the weather in the interests of impoverishing bona fide road-tax-paid travellers?

The climate of the UK, if it has one, has always been variable, driven by opposing winds and currents, the Gulf Stream one minute, an Arctic blast the next. We remember summers of torrential rain, hail stones like ping-pong balls, then balmy winter evenings when the sun shone gently on its long suffering, typically British 'musn't grumble' subjects. My triple-graduated magna-cum-laude niece now boasts about cycling to work, without mentioning how she could have done so with a toddler, a baby and three bags of shopping as a young mother. Thankfully, after a public backlash, her local council has been forced to abandon its illegal congestion charge.

We need only check the paintings of Peter Breugel: 'The Hunters in the Snow' 1565, frost fairs on the Thames and elsewhere in Europe in the 17th and 18th centuries or 'The Harvesters', to know that extremes of temperature were nothing new until the beginning of the end of the Little Ice Age in the mid 19th century, long before our so-called 'carbon footprint' was adopted as a measure of global warming in the anthropogenic climate change cult. Cherry picking data is the favourite pastime of globalist 'experts'. Awkward data that contradicts the supposed 'science'—surely, the most abused word in the English language—is conveniently ignored. The infamous 'hockey stick graph' (1988), for example, purported to show the average annual temperature in the Northern Hemisphere from 1400 A.D. followed by a sharp upward trend in the 20th century attributed to man-made industrial activity. Unfortunately, real investigative scientists have pointed out that significant years were omitted from the graph showing a further downward trend.[1]

Ever keen to check things out for myself, I attended an official International Panel on Climate Change (IPCC) public meeting in our local, self-proclaimed prestigious conference centre. Free wine was served to the attendees, all seemed terribly civilised and I was looking forward to a lively discussion. The speakers included a representative from two local universities, the local airport as well as the IPCC itself. After the presentations, questions were invited from the audience but although I had my hand up, the microphone never

1 CDN, 'IPCC Pressure Tactics Exposed: A Climategate Backgrounder', 11 June 2023, *Principia Scientific [website]*, <https://principia-scientific.com/ipcc-pressure-tactics-exposed-a-climategate-backgrounder>, accessed 22 August 2024.

seemed to stray in my direction. Eventually, I stood up and asked my pointed questions relating to scientific evidence for climate change being caused by human activity rather than computer models. At the end of the meeting I was taken aside by the chairwomen, much as one would consign a recalcitrant child to the naughty step and given a good scolding for challenging the 'science', as if real science was ever settled. Later it emerged that the chief executive of the Conference Centre was awarded a bonus of over $72,000—quite irrelevant information, I am sure.

In reality, I had already read my Schellenberger, Moore, Crockford, Happer, Lindzen, Keenan and many others from the 1,500 or so real scientists who have stated the truth about the narrative—for that's all it is—much like the narrative that brainwashed whole populations into believing the one about a dangerous pandemic that never actually happened, while our World Economic Forum puppet politicians betrayed their own disbelief in it (see Boris Johnson and chums at Partygate). Do as I say, not as I do sprang to mind in those now far-off days where the first casualty of war is truth. An amusing cartoon depicted how the Amish had managed to avoid Covid-19 infection, they replied 'We do not watch TV'. Mainstream media are a vital tool in the information war, keeping up the catastrophising of natural events and instilling fear into populations which then become easier to control.

So what's behind all this fearmongering, Net Zero and CO_2 propaganda? Let's first brush up on what we already have deduced as reality. We know that carbon dioxide (CO_2) is essential to life and plants. (Every time Prince Charles spoke to his plants, he was releasing CO_2 in his breath, which is absorbed through their leaves). CO_2 keeps the planet's temperature stable in order to sustain life. Since it began, planet earth has undergone periods of climate change—some warm, some colder periods but this is mainly the result of solar activity and possible volcanic eruptions plus the oceans which release some CO_2—an invisible gas which is only 1% of so-called greenhouse gases.

The Great Barrier Reef, the coral, the Arctic ice and the polar bears are just fine but notice how the fearmongers target those phenomena which are quite difficult for you to check! Temperature increases cause CO_2 to rise—not the other way round. Grass-fed animal farming does not affect CO_2, on the contrary it benefits the soil as well as your health. Factory farms on the other hand emit toxic antibiotic residues and growth hormones into the water supply. Dangerous weedkillers such as Roundup (Glyphosate) deplete the soil and are suspected of causing chronic diseases such as cancers, diabetes and cognitive decline. The long-term effect of genetically modified crops, along with the increasing coverage of 5G, may have unknown consequences to human health.

The fossil fuel industries have brought unprecedented wealth to your average citizen. Remember Harold Macmillan's 1957 'You have never had it so good.' He was right in the sense that we had decent council homes with gardens, electricity and running water, sanitation and a free health service. Electric cars rely on expensively produced batteries and substances extracted from mines probably by child labour. Wind farms are disastrous to wildlife and inefficient when the wind does not blow. Heat pumps replacing cheap to run gas boilers are expensive to instal. Oil is a natural resource not likely to run out soon. It's true that our mainly 19th century built cities and towns were not designed with the family car in mind, but it's not the family car that is holding up traffic so much as the endless roadworks, diversions and closure of city side streets by local Councils that make the problem worse, not better. 15-minute cities, smart cities or indeed smart anything, cashless societies, etc. are methods of controlling the population with the ultimate aim of reducing it. The fake climate agenda is just another method of impoverishing ordinary people and corralling them into zones, while controlling their food and money supply.

Professor William Happer is right in suspecting that our neo-feudalist future was planned long ago by the military industrial complex, the technocrats, the banking fraternity and assorted billionaires who acquire power through money and corrupt influence. Families such as the Rockefellers and Bilderbergs, the Trilateral Commission and the banking fraternity have been in control for decades. Non-governmental organisations such as the WHO, the UN, NATO, the IMF, the EU, the WEF were long ago infiltrated by the globalist clique who have been directing world affairs for a very long time with the help of their puppets in elected governments. They represent only 1% of the population, so the 99% of the remainder must fight back and refuse to comply. 'If voting in elections made any difference, they would not let you do it.'

We won't be lectured by celebrities on the dangers of rising CO2 levels as they fly around the world on private jets telling us how we should do our bit to reduce our carbon footprints.

—*Liam Edwards*

XXIII.

It's hot because it's cold

By Liam Edwards

> Liam is a writer/ poet from the West Midlands. He began writing to try and make sense of these strange times we find ourselves in, with the hope of finding some answers to the problems we face today. He believes that when it comes to fighting injustice, the pen is mightier than the sword and that through words, we can shine the light of truth into the face of darkness and into the minds of others.

As I sit looking out the window on this grey July morning the only thing I see that concerns me is the unnatural smoke-like blanket of grey cloud that never seems to clear, yet I am told we are in the midst of a climate emergency; the earth is getting too hot they say. My phone tells me it's currently a scorching 14 degrees, it's so hot in fact that I've had the heating on, yet if I'm to believe what I've been told I am lucky to be alive, as according to Swedish climate messiah Greta Thunberg in a now deleted 2018 doomsday tweet, we shouldn't even be here, all humanity should have been wiped out by 2023. How could Greta, the cunning meteorologist that she is, have gotten this so wrong? Could the science that is driving this emergency be wrong? No, it couldn't be, what about the images I've seen of the sad lonely polar bears lost adrift on tiny pieces of ice sheet on the news? Surely not. This *is* a climate emergency! We are on the brink of extinction! Or so we are told. But consider for a moment that maybe the only 'Climate Emergency' we are facing is actually a deeply corrupt 'Political Climate' and that we are again being sold a lie as they attempt another power grab in the name of public safety, another coalition between world leaders, banks and corporations to seize more money and powers in the form of another 'Global Emergency' controlling food supplies, restricting travel and energy use until we are all sat in darkness praying next door's Tesla spontaneously combusts so we might warm our frostbitten hands.

They forget one thing, however, this is not our first rodeo; we are battle tested and ready to fight for our freedoms. We won't be lectured by celebrities on the dangers of rising CO_2 levels as they fly around the world on private jets telling us how we should do our bit to reduce our carbon footprints. Billionaires giving speeches about Net Zero while launching rockets into space emitting 1,000s of tons of CO_2 into the atmosphere, giant corporations and airlines which produce more CO_2 than small countries pretending they care about the environment by 'offsetting' their emissions and buying up bogus carbon tokens, 95% of which, Verra, the world's largest carbon standard agency say are completely fraudulent[*] and thus washing their hands of any accountability, promising to plant hypothetical trees and protect indigenous land while simultaneously destroying it. This is the insane logic we are seeing coming from the climate cult; it's

---

[*] A nine-month investigation undertaken by *The Guardian*, the German weekly *Die Zeit* and *SourceMaterial* revealed that 90% of (Verra) rainforest offset credits are likely to be 'phantom credits' and do not represent genuine carbon reductions. *(ID)*

See: Patrick Greenfield, 'Revealed: more than 90% of rainforest carbon offsets by biggest certifier are worthless, analysis shows', 18 January 2023, *The Guardian* [website] <https://www.theguardian.com/environment/2023/jan/18/revealed-forest-carbon-offsets-biggest-provider-worthless-verra-aoe>, accessed 14 September 2024.

like me saying I'm going to become a serial killer but it's ok, because I'm also going to sign up to become an organ donor because I care about life.

We are being told by governments that we must switch to dangerous 'sustainable' electric vehicles made from the blood and sweat of child slaves in cobalt and lithium mines which produce millions of tons of toxic waste contaminating the air for hundreds of years. The same governments that are conducting dangerous geoengineering projects such as SRM (Solar Radiation Management), aka 'chemtrails' spraying poisonous sulphates into the stratosphere. Playing god without knowing the long-term effects on the environment and the risks they pose to our health. Forcing poor countries to focus on expensive renewable energy, denying them access to their own natural resources, the same resources they themselves seek to exploit and control and destroying their hopes for future development and a chance at becoming independent. They are going after farmers, raising prices of fertilizers and feeds, ordering them to cull their herds by 95%, essentially putting them out of business in an attempt to control food supplies, even telling us what we can and can't eat. But how long will it be until people get tired of their bug rations and the rich start looking like juicy and well marbled alternatives.

Once again the proposals being put forward to combat the 'Climate Emergency' such as carbon taxes will only affect the average hard working citizens of the world; those who can't buy their way out of responsibility will suffer the most. The big corporations on the other hand, will combat this by raising their prices to recoup any losses. Effectively pricing us out of things like travel, making it a luxury only affordable to the super rich. All the while remaining business as usual for the so-called 'Elites'; this is the hypocrisy and depravity of the green agenda. Governments and corporations pretending to care about the earth while continuing to profit from its destruction, using climate change as a guise for more nefarious plans such as globalisation.

Now, I am by no means what they would call a 'climate change denier' a ridiculous term if you think about it, for I do not deny there is a climate, nor do I deny that it is changing. But history shows us that the climate changes with or without the help of humans and if we are to believe CO_2 is the main driving factor in climate change, causing earth's temperature to rise to dangerous levels not previously seen, how do they explain that CO_2 was actually higher during the ice age? Or that the earth was actually 0.7 degrees warmer 6,000 years ago? Are we to believe it's actually hot because it's cold? The truth is despite all the money pumped into funding for research, there is still no scientific proof that man-made CO_2 is the driving force in climate change. A fact the president of the 2023 United Nations Climate Change Conference (COP28) Dr Sultan Al Jaber also pointed out during his speech to the committee in Abu Dabi. Dr Sultan is also the chair of Abu Dhabi National Oil, responsible for pumping 2.7 million barrels of oil a day, a figure he hopes to double by 2027, an interesting choice of president and host nation for the COP28 considering they didn't even sign the agreement.

You see the CO_2 we produce accounts for less than 5% of greenhouse gases, volcanoes produce more CO_2 than all the factories, planes and cars combined, what do they suggest we do about those pesky angry fire mountains? For one you can't tax a volcano, I'm surprised they haven't suggested making them wear giant surgical masks, we all know how well they work and I'm sure there are plenty still knocking around. There is also evidence to the contrary from leading meteorologists on the study of sunspots that suggests it is actually solar activity driving earth's temperature, not CO_2; imagine that, the giant ball of burning gas having an effect on temperature? Who'd of thought it! You can't simply ask people to trust the 'science' and ask us to completely change our way of life without first presenting all the facts. That in itself is unscientific.

I admit I am no expert on climate change, but I know when I smell a rat, there are too many inconsistencies in the narrative, inconsistencies they refuse to address. Why do they seek to discredit and silence anyone who tries to draw attention to the elephant in the room? We are seeing the same psychological warfare used during the 'pandemic' to demonise and ridicule anyone who speaks out against the rhetoric being pushed by Legacy Media. But this time instead of being asked to give up our freedoms to protect the elderly and vulnerable, it's now to protect mankind itself from extinction and if for any reason you choose not to agree with that 'science' you must be one of those conspiracy theorists and want everyone to die, how selfish.

No doubt the biggest question from those who accept the whole CO_2 climate disaster theory will be why would they lie about all this? Why indeed, the problem is they are just not asking the right question, the question should be what's in it for them? Simply follow the money and that will give them the answers they are looking for. The climate change industry is now worth trillions worldwide. That's a lot of money to just walk away from, and one hell of a reason to justify a threat that perhaps isn't as bad as first thought. Institutions built on the back of funding to prove the correlation between CO_2 and climate change would simply cease to exist overnight, thousands of jobs and careers would be on the line and an entire branch of science would suddenly become obsolete. Food for thought.

Now I don't doubt there are a lot of good people out there who genuinely want to make the world a better place, but there are also governments and corporations that may seek to exploit such a situation and I can think of two big reasons. What are two things they can never get enough of? Money, and Power, and if they see a chance for more, be sure they will take it. Another important question people should be asking is do you actually trust the government? If you had someone in your life who repeatedly lied to you making your life miserable and knowing the lies they told ultimately cost lives, would you trust them? If they lied about one thing, be sure they will lie about anything. These are dangerous times we live in, with a new form of 'climate extremism' at the forefront, democratic debate is no longer allowed and we are again seeing the censorship of any information that could threaten the narrative. This is why it is vital we research these issues for ourselves. Mainstream media has proven it is just a paid appendage of the corrupt globalists and has lost all credibility when it comes to reporting the truth. We know from experience it's useless trying to change the minds of those who don't want to listen, but we must keep sewing the seeds of doubt, which will inevitably bloom within the subconscious causing people to question what they are being told.

I believe the biggest threat the Earth is facing today is not climate change, but greed, the same greed that threatens to wipe out entire species of plants, insects and animals through deforestation and the use of dangerous pesticides, the same greed responsible for killing off marine life by polluting our seas and oceans and the thousands of innocent people that die every year in illegal wars fought over the same resources we are told we shouldn't be using. I wonder what the actual carbon footprint of the military industrial complex could be? An unofficial figure suggests 5.5% but you can be sure it's much higher as the COP28 committee does not require them to disclose such figures and conveniently makes no mention of military and conflict emissions in the agreement. Maybe this would be a better place to start on their road towards Net Zero instead of making us feel guilty for enjoying a steak dinner or heading off on a family holiday abroad. I suggest we formulate a Net Zero strategy to rid the world of the parasitic entities responsible for decimating our planet. Humanity stands at a precipice, we must unite against the evil of this world, a silent killer more deadly than any greenhouse gas. It's clear the 'climate agenda' is just another 'trojan horse' painted 'green', but there is still hope for the future. We all know in our hearts there is a better way than this madness they call 'Normality'. This is a time of mass awakening, a rise in global protests show momentum is building, more people are saying no to gift wrapped austerity in the name of public safety. The power lies with us, and together we can build a future where truth prevails and natural order is restored, and then, and only then, will the Earth begin to heal.

To my knowledge, the IPCC does not address extreme weather events and have never used the words climate crisis. The so-called climate crisis is a fiction coined by the press, politicians, and extremist groups. But when the world's news media and politicians use this term, they have not been corrected by the UN or the IPCC. This is all part of the propaganda machine to nudge us, to scare us and to convince us that we will do anything to reverse this 'crisis'.

—*Martin Chambers*

XXIV.

It's all about power

By Martin Chambers

I do not want or need to address the science, or lack of, surrounding climate change. My focus is on the power. With power comes money and in today's world, power is everything. Why is it so important for politicians, scientists, journalists to promote climate change? The answer is power, power and money.

When it comes to the science, we hear the term 'settled science' which means 'don't argue'. Any dissenting voices are dismissed as 'climate deniers' and accused of spreading misinformation, pilloried, and cancelled. Question the science at your peril. I believe that most of the scientists that have spoken out against climate change are retired from their careers and have very little to lose from speaking out. All the others (by exclusion) are in paid employment. Their employers are universities and as well as other organisations set up to promote climate change and, more importantly, the remedies to deal with it. It is well known that any scientist or researcher who dares to step out of line will quickly lose their job or funding. The main reason for this is that these institutions are funded by the very people and organisations whose agenda is at work here. Some call it philanthropy; I call it buying science. And buying science has been happening for years, whether it is the tobacco industry, of Big Food or Big

Pharma, or the NHS. It is all the same. The people with money can influence the outcomes of research papers and scientific studies and do so on a continual basis. We saw it massively at the start of the 2020 with Imperial College London and Neil Ferguson's predictions.

I used to trust journalists. I trusted them to tell me the truth and even better, to tell me that this or that is an untruth or worse, a lie. I trusted them to cross-examine a scientist or a politician to try to establish the truth, to look deeper into the data, to ask awkward questions, to establish what is behind the rhetoric, to accuse them of having an ulterior motive. While that trust was still there to some extent in March 2020, it disappeared then. This is the greatest crime for the once noble profession of journalism where souls have been sold to the devil of state-owned and state-run media.

In 2018, Fran Unsworth, the BBC's director of news and current affairs at the time issued internal guidance to journalists on how to report climate change. She added: 'With this in mind, we are offering all editorial staff new training for reporting on climate change. The one-hour course covers the latest science, policy, research, and misconceptions to challenge, giving you confidence to cover the topic accurately and knowledgeably.' So as far as BBC News is concerned, there is no debate.

It is one thing to hide the truth, it is a step further to tell outright lies. The UN's Intergovernmental Panel on Climate Change (IPCC) has made some dramatic predictions about the changing climate in years to come. In discussing the past however, they are more dependable particularly when it comes to extreme weather events of which, evidence says, there have been fewer over the years. But the BBC and Sky News, as well as other outlets disagree. They continue to report such news items as if they are increasing year-on-year. In June this year the BBC posted a news item with the headline

'How climate change worsens heatwaves, droughts, wildfires and floods'. They augment this with red weather maps when the sun shines, but it is not just the BBC, they are all at it. The 'it' is alarmist propaganda peddled for one reason only, to scare us. 'We will all die unless we install a heat pump or buy an electric car'. It began in 2020 when we had the wits scared out of us, all of us. Someone once said, 'kill one man, frighten ten thousand'.

To my knowledge, the IPCC does not address extreme weather events and have never used the words climate crisis. The so-called climate crisis is a fiction coined by the press, politicians, and extremist groups. But when the world's news media and politicians use this term, they have not been corrected by the UN or the IPCC. This is all part of the propaganda machine to nudge us, to scare us and to convince us that we will do anything to reverse this 'crisis'.

I also believe that the career-limiting aspect to speaking out amongst journalists is a factor that strangles their otherwise enquiring mind and their powers of reasoning. I am aware that I am painting all journalists with the same brush and that there are many individual heroes that do speak out and I thank them for this, but they are not listened to by the mainstream media. We need more of these, and we need to make them our first choice for honest and factual news and comment.

I would like to ask where all the socialists have gone? Where are the champions of the working class, the poverty stricken, the single parents, the unemployed, the low paid, the unskilled? I just do not see them; they must be there, but they are keeping very quiet. Mr Starmer talks about change, but I struggle to see the difference between his party and the outgoing party. The two of them are a uni-party, as someone said recently, two cheeks of the same rear-end. The most significant demonstration of this is their commitment towards

Net Zero.

Mr Miliband gave us the 2008 Climate Change Act, which has already cost the country a fortune. Subsidies for renewable energy mandated by the Act have already added more than £100 billion to our energy bills. These are the energy bills paid for by people that used to be looked after and cared for by socialists but now they do not care at all. They do not figure in any manifesto, they are not mentioned at all, we have a socialist party in name only.

Thanks to Mr Miliband's obsession with wind and solar power, the annual subsidy bill will have risen to £30 billion in five years if Labour's plans are carried out. There are continual requests from wind farm companies for additional subsidies when the wind does not blow. We pay for these.

The costs imposed do not stop there. Labour have already said they will ban the sale of new petrol and diesel cars by 2030. This is nothing to do with carbon dioxide and everything to do with getting us off the road and out of our cars. This is what 15-minute cities and low emission zones are all about. It is all part of the plan.

This will cost the country over a trillion by 2040. This is our money, paid in taxes to the government. What then for the single parent or the delivery driver. I do not believe we voted for this. I do not believe this is democracy.

If I were asked by one of my children for career advice, I would say 'get into climate change'. OK, from where I stand on this, my advice would be a little flippant but make no doubt, this is where the money is: central government, local government, the civil service, NGOs, consultancy firms, the list is endless. This is why challenging the science has no effect on these people. That does not mean we should stop pushing the scientific arguments particularly at the local level. But it is a huge business with its own momentum, it is unstoppable. And the reason for this is that too many people are making too much money out of it.

I am unable to ignore the geopolitical aspects to this. Climate change, as an idea, was invented by the Club of Rome in the 1960s, who said 'The common enemy of humanity is man. In searching for a new enemy to unite us, we came up with the idea that pollution, the threat of global warming, water shortages, famine and the like would fit the bill.' If you look at the membership of the Club of Rome, it is filled with the rich and the super-rich. It was then and it is now. Then you add in the World Economic Forum, the Bilderberg Group and the Council on Foreign Relations, the Clubs of Madrid, Vienna, and Budapest too. None of the members of any of these groups are about to argue against climate change because that would not be good for business. These organisations do not get reported on by the BBC or any other mainstream news organisation. This is a playbook where the rules have been set by the rich and powerful. The banks and the corporations weald their influence in a seamless web of control that impacts everyone.

Moreover, this has escalated since the pandemic, the political pressure has increased, climate is now on the agenda for the WHO, the censorship has been ramped up, the government is buying huge tracts of farmland, soon meat will be banned. There is no doubt in my mind that the pandemic has given them licence to ramp up the propaganda and fear messaging. You could argue that the pandemic was a rehearsal for climate lockdowns.

We can fight this, but it will need a lot of effort. We must not be cowed, we must not give up—that is what they want, that is the way of the slave. I believe we can have influence by electing as many independent MPs as possible. We already know that the uni-party system is undemocratic. In fact, any political party is part of the problem and in this way, the Reform Party is a Trojan Horse (if they want a hero, we will give them one).

Finally, there are three main battlegrounds here, three prongs to the devil's trident: Climate, Health and Conflict. All three with an ulterior motive to frighten us, control us, starve and kill us. My awakening came when I realised that my MP was just a pawn, that my government is not there for me, that the NHS is already privatised, that the BBC is an arm of the state, that the pandemic was planned, that the banks' real intent is to steal everything we own, that NATO is simply there to crush Russia, that the WHO wants to destroy the nuclear family, that the Royal Family are in it up to their necks.

I believe we can defeat them, but we must know them first. It is by doing our research that we achieve this.

There is a gulf between real science and the environmentalist agenda. It takes only rudimentary general scientific knowledge to trip up the very poor arguments at the first hurdle and only a very small amount of commonsensical clear thinking. When people are afraid, poor decisions are made and the promoted narrative is accepted without question. Because of fear.

—*Louis Brothnias*

XXV.

Anthropogenic climate change— scientific fraud, part three

By Louis Brothnias

Earth's moon and the continuous movement of the global seas (tidal motion) offers the source of endless free energy. The only reason to shun the whole concept is the lack of opportunity to make a very, very substantial financial return (trillions). That hydrogen would make a good candidate is only because it is assumed to be present everywhere in the Universe. How this source would be 'mined' or the gas liquefied and stored is never discussed but simply referred to as Green Hydrogen as though that makes it superior to ordinary hydrogen. It is currently produced using oil-based fuels. If generated electrolytically from water, electricity is an essential requirement. The gargantuan dangers involved with such a liquefied fuel are never mentioned. What might be the outcome in the event of a single vehicle crash or fire? Or a tanker delivering liquid hydrogen. And sitting above a tank of liquefied hydrogen in a car would be particularly unnerving. Only if the challenges of hydrogen gas liquefaction, storage, and convincing users of the safety of using hydrogen as a fuel were to be overcome, could there potentially be any achievable financial return. Nevertheless, the consumption of hydrogen in a fuel cell will consume O_2 and convert it into water (H_2O). The world-wide production of new water from hydrogen-cell fuelled vehicles will increase global warming. The very thing that the hydrogen fuel cell is claimed to reduce by not producing any CO_2. The narrative would have it that carbon dioxide is a toxic gas but fails to point out that it is the very gas that will generate new O_2 (photosynthesis). Deforestation is a genocidal (long-term) plan. Less trees means that less CO_2 will be removed from the atmosphere and less O_2 will be created to sustain an ever-growing population. The CO_2 removed by trees from the atmosphere to produce O_2 and glucose is the same amount of CO_2 that will be formed when it is consumed. Similarly, the water (H_2O) used is simply regenerated. There is no nett change in CO_2 or H_2O but hydrogen fuelled vehicles produce new water and consume oxygen. Direct competition will occur between the hydrogen fuelled vehicle and all life for the available oxygen. The introduction of the hydrogen fuelled vehicle is a genocidal plan.

Photosynthesis: $\quad 6CO_2 + 6H_2O \rightarrow 6O_2 + C_6H_{12}O_6$ (glucose)

Glucose combustion: $\quad C_6H_{12}O_6 + 6O_2 \rightarrow 6CO_2 + 6H_2O$

Hydrogen cell: $\quad 2H_2 + O_2 \rightarrow 2H_2O$ (new water)

There is a gulf between real (good) science and the environmentalist (bad science) agenda. And it takes only rudimentary general scientific knowledge to trip up the very poor arguments at the first hurdle. And only a very small amount of commonsensical clear thinking. When people are afraid, poor decisions are made and the promoted narrative is accepted without question. Because of *fear*.

Consider a World where oil-based fuels were not used. Transporting anything around the planet would not be realistic. A family-size car fuel tank will hold around 10–12 gallons (US/imperial) and there are about 1.5 billion vehicles around the World. Theoretically, the maximum volume of fuel would be 15–18 billion gallons at any one time. Thousands of merchant ships (58,228) were operating worldwide in 2022. The largest container ship can hold 4.5 million gallons of fuel oil. Just this one ship equates to around 450,000 cars. It would need only 4,000 such container ships to carry as much or more fuel as all the vehicles on Earth. There are 5,574 container ships of various sizes and would consume at least as much or more fuel (fuel oil) as all the fuel (petrol/diesel) used by the 1.5 billion cars (15–18 billion gallons) from around the World. Heavier fuel oils (shipping) are far more polluting than the more refined and more volatile petrol and diesel. Around 80% of goods are transported by ship.

Merchant ships (58,228)	
Ro–Ro/General cargo:	17,784
Bulk cargo carriers:	12,941
Crude-oil tankers:	8,258
Chemical tankers:	6,122
Container ships:	5,574
Passenger ships:	5,369
Liquid natural gas tankers:	2,180

Electrically powered aircraft, road, and rail transport would make up the other 20%. The weight of an aircraft battery (even if technically achievable) would be enormous. A car battery weighs ~500kg (25–30% curb weight). The elephant in the room, however, is the production of the electricity to recharge the batteries. It is an argument that becomes totally absurd. How the electricity would be generated is never discussed. It could only be provided in the necessary enormous quantities by using oil-based fuels. The entire fraud is completely lacking any logic and is moronic thinking. Cars and other forms of transport will all become obsolete as the resources required to enable it quickly run out. It's all part of the plan to make all cars obsolete and imprison the entire global population. The fraud is fraught with deceit and lies and is an existential threat to all life on Earth. In the attempt to introduce any kind of measure or new technology to fix a climate system that is not damaged and a planet that doesn't need saving, the environment and the climate will be destabilised and severely and detrimentally changed. Adding water to promote atmospheric warming and technology to remove life-sustaining O_2. What is the alleged irreversible damage? This question has never even been asked. It is unanswerable.

No farmers no food.

The Green Deal is a death deal. Without us you don't eat.
Livestock farming died today.

—*Protesting farmers in Europe*

XXVI.

Repent your sins for the end of the world is nigh

By Aethel

> Aethel is a retired English and languages teacher. She also trained as a nutritionist after recovering from health problems. She has worked in England and for the Papua Ekalesia, Papua New Guinea's first self-governing church.

No, this isn't from one of those cranks parading the streets with billboards, the ones you used to see in days of yore. Nobody believed them and the end of the world never happened. No, this is a government warning. It's official, quite different: we have to change our ways. If we don't it will get hotter and hotter and we'll all boil over.

Government mouthpiece, Greta Thunberg, predicted we'd burn up last year. Remember all the, 'Come on, chop chop, get a move on? If we don't step on the gas (turn off the gas?) we're done for. Only a few months left before Armageddon'.[1]

It didn't happen, and it doesn't look like it will happen this year either, despite the Met Office claiming May was the

---

1   —Steve Forbes, 'The Case Of Greta Thunberg's Deleted Tweet—What Alarmists Need To Hear', 14 July 2023, *Forbes* [website], <https://forbes.com/sites/steveforbes/2023/07/14/the-case-of-greta-thunbergs-deleted-tweet---what-alarmists-need-to-hear>, accessed 21 August 2024.

hottest on record.[2] (Where do they put their thermometers?) It's been so cold my daughter wore her winter coat right up till June, I was still in my winter clothes and when I went to buy a summer coat there weren't any. All you could get were cold-weather padded ones. The UN Head of Climate Change has updated the forecast.[3] There's not a shadow of doubt, the end of the world is coming, he says. If we don't follow Net Zero to the letter that is.

Official reports state that 98.7% of scientists believe in climate change.[4] Of course, believing isn't proof, but they wouldn't say that if it wasn't true, would they?

Just think of the effort required to arrive at that figure. Did they have any help? There are over eight and a half million scientists in the world. How many special advisers and translators did it take to contact them all? Did they use letters or emails? Something like this perhaps?

Sign with an X:

I believe in climate change.

I do not believe in climate change.

Your university will lose funding and you will lose your

2 —PA Media, 'May and spring were warmest on record in UK, Met Office says', 3 June 2024, *The Guardian* [website], <https://theguardian.com/uk-news/article/2024/jun/03/may-spring-warmest-record-uk-met-office>, accessed 21 August 2024.

3 —Alexandria Williams, 'UN Secretary-General issues wakeup call on climate change', 6 June 2024, *Deutsche Welle (DW)* [website], <https://dw.com/en/un-secretary-general-issues-wakeup-call-on-climate-change/video-69291009>, accessed 29 August 2024.

4 —Krista F Myers *et al* 2021 *Environ. Res. Lett.* 16 104030; 'Consensus revisited: quantifying scientific agreement on climate change and climate expertise among Earth scientists 10 years later', 20 October 2021, *Institute of Physics: IOPScience* [website], <https://iopscience.iop.org/article/10.1088/1748-9326/ac2774>, [file], <https://iopscience.iop.org/article/10.1088/1748-9326/ac2774/pdf>, accessed 21 August 2024.
—Earl J. Ritchie, 'Fact Checking The Claim Of 97% Consensus On Anthropogenic Climate Change', 14 December 2016, *Forbes* [website], <https://forbes.com/sites/uhenergy/2016/12/14/fact-checking-the-97-consensus-on-anthropogenic-climate-change/>, accessed 29 August 2024.

job if your X is put in the wrong place.

Then there's all the time it would take, looking up the addresses, dispatching the missives. Kabul University—I know there's one there—and in Beijing and other parts of China. Inner and Outer Mongolia, how many universities do they have? Or Tajikistan, Turkmenistan, Uzbekistan, places like that? And there are all the firms that use scientists, research projects, councils, retired scientists. Anyone else? Did they miss any? Did they get a 100% response?

Well, actually folks, 10,929 invites were sent to geoscience institutions asking if they'd partake in the survey. That's all. And guess how many replies they received. 2,780! After comparing the answers with other published scientists, 153 experts were selected and 98.7% of them agreed we're making the planet hotter.

In contrast, 1,936 distinguished scientists from the Global Climate Intelligence Group, including two Nobel Prize winners, have signed a declaration stating, 'There is no climate emergency'.[5] You have to hand it to them. It's brave to stand up to the Government like that.

Archaeologists tell us the Romans grew grapes along Hadrian's Wall.[6] The variety used was probably Bacchus, and in Southern Scotland Vitis Vinifera was grown. Today the furthest north that's warm enough for grapes is Yorkshire. I'm told there are a number of varieties and they make beautiful wine there, absolutely first rate. But further up? Have you ever been to Northumberland? Do you know what it's

---

5 —Global Climate Intelligence Group, 'World Climate Declaration. There is no climate emergency. A global network of over 1400 scientists professionals prepared this urgent message.', *CLINTEL* [website], <https://clintel.org/world-climate-declaration>, 16 July 2024 [file] <https://clintel.org/wp-content/uploads/2024/07/WCD-240716.pdf>, accessed 21 August 2024.

6 —Ron Smith, 'Climate change and the growing of the grape', 14 April 2013, *Breaking Views* [website], <https://breakingviewsnz.blogspot.com/2013/04/ron-smith-climate-change-and-growing-of.html>, accessed 29 August 2024.

like? A beautiful, beautiful county, but far too cold for grapes. How come then, if it was so much warmer in Roman times, they didn't boil over and go pop?

George Monbiot has written a number of articles in the Daily Mail and the Guardian claiming there's no point in having farmers anymore.[7] Their animals and practices are detrimental to the planet; we must eat bugs and synthetic food instead. He has written a book about it, *Regenesis*, and toured Europe promoting his ideas.[8] Well, they're not really his, they're government ideas, all governments endorse them and they were originally dreamed up by Klaus Schwab, the man who makes all our laws. No more meat. No more lovely animals in the fields. Can't they think of any alternative strategies?

Farmers are up in arms about it. Absolutely steaming with rage.[9] Things have been made so difficult for them (to get rid of them of course) and they've been kicking up a fuss in England, Ireland, Wales, all over France, all over Germany, in Belgium, Spain, Portugal, the Netherlands, Italy, Greece,

7   —James Robinson, 'George Monbiot says farming should be ABOLISHED to save the planet: Climate change activist says meat can be replaced with lab-grown food like protein pancakes', 18 May 2022, *Daily Mail* [website], <https://www.dailymail.co.uk/news/article-10828875/George-Monbiot-says-farming-ABOLISHED-save-planet.html>, accessed 29 August 2024.
    —George Monbiot, 'Lab-grown meat will soon destroy farming—and save the planet', 8 January 2020, *The Guardian* [website], <https://theguardian.com/commentisfree/2020/jan/08/lab-grown-food-destroy-farming-save-planet>, accessed 29 August 2024.
    —George Monbiot, 'Goodbye—and good riddance—to livestock farming', 4 October 2017, *The Guardian* [website], <https://theguardian.com/commentisfree/2017/oct/04/livestock-farming-artificial-meat-industry-animals>, accessed 29 August 2024.
8   —George Monbiot, *Regenesis: Feeding the World without Devouring the Planet*, 2022.
9   —Cécile Boutelet *et al*, 'Farmers' anger is mounting across Europe', 24 January 2024, *Le Monde* [website], <https://lemonde.fr/en/economy/article/2024/01/24/farmers-anger-is-mounting-across-europe_6460470_19.html>, accessed 21 August 2024.

Poland, Romania, Lithuania, the Czech Republic, and some other European countries, blocking the streets with their tractors, dumping manure outside officials' houses and waving flags saying 'No farmers, no food'.

The Greek farmers went even further.[10] They waved black flags saying 'The green deal is a death deal. Without us you don't eat. Livestock farming died today.' And they accompanied them with a coffin and wreaths.

You can understand it, can't you. Farms and farmland held by families for years, generation after generation, all to be cruelly snatched away from them. In the Netherlands, can you believe it, they were going to build flats for immigrants on the land instead.[11] 200,000 cattle to be culled in the Irish Republic,[12] Danish farmers to be charged £80.00 a cow,[13] fuel taxes hiked, unwarranted overloads of paperwork.[14]

In Brussels farmers threw eggs and stones at the

10 —Derek Gatopoulos Associated Press, 'Greek farmers bring black-flag protest to Athens', Athens, Greece, 5 March 2013, *Associated Press News (AP News)* [website], <https://apnews.com/article/dc770cde069c4f7bba9406e7ca8f76f8>, accessed 21 August 2024.

11 —Frank Bergman, 'Dutch Farmers Topple Agriculture Minister Leading Radical Climate Agenda', 6 September 2024, *Slay News* [website], <https://slaynews.com>, accessed 29 August 2024.

12 —Jude Webber, 'Irish farmers pressured to cull up to 200,000 cows to meet climate goals', 10 August 2023, *Financial Times* [website], <https://ft.com/content/4028ae15-fea2-48fb-bcdb-228f61e1b098>, accessed 29 August 2024.

13 —Emma Gatten, 'Denmark to charge farmers £80 per cow in world-first meat tax', 26 June 2024, *The Telegraph* [website], <https://www.telegraph.co.uk/world-news/2024/06/26/denmark-charge-farmers-per-cow-in-world-first-meat-tax>. —Emma Gatten, 'Denmark to charge farmers £80 per cow in world-first meat tax', 27 June 2024, *Meatex* [website], <https://meatex.co.uk/2024/06/27/denmark-to-charge-farmers-80-per-cow-in-world-first-meat-tax>, accessed 29 August 2024.

14 —Euronews Green with APTN, '"With the smallest error, there are fines": EU farmers' frustration grows as paperwork piles up', 15 February 2024, *Euro News* [website], <https://euronews.com/green/2024/02/15/with-the-smallest-error-there-are-fines-eu-farmers-frustration-grows-as-paperwork-piles-up>, accessed 29 August 2024.

European Parliament. They started fires and set off fireworks as they demanded more help.

'We want to stop these crazy laws that come every single day from the European Commission,' said Jose Maria Castilla, a farmer representing Spanish farmers' union Asaja.[15]

It's tragic, but governments insist if we don't make sacrifices the planet is doomed. It's all in Net Zero: no more meat, fruit and veg to follow, then it will be bugs and synthetic foods for evermore. Bugs are already being included in school dinners to get children used to them.[16][17]

The trouble is, as a qualified nutritionist, I know they're not good for us. Some time back I signed a petition for the Soil Association asking the government not to promote ultra-processed food. The Soil Association says it's not good for people too.

Our digestive systems have evolved over thousands of years to cope with the foods growing around us. They don't

15 —Reuters, 'Farmers' anger spreads in Europe, governments promise help', 2 February 2024, *The Economic Times* [website], <https://economictimes.indiatimes.com/news/international/world-news/farmers-anger-spreads-in-europe-governments-promise-help/articleshow/107330922.cms>, accessed 29 August 2024.<https://m.economictimes.com>

16 —Rachael Davies, 'In an effort to be more sustainable, elementary schools are serving insects during lunch: "These issues are important"', 3 January 2023 (Elementary schools in Wales are serving their kids insects: 'Pupils at four primary schools in Wales will be trying out a new source of eco-friendly protein: insects.'), *The Cool Down* [website], <https://thecooldown.com/sustainable-food/insects-lunch-schools-wales-bugs-protein>, accessed 29 August 2024.
—Lorelli Mojica, 'School trialling insects on the menu to support sustainability, but is this the way forward?' 16 May 2023, *Twinkl Digest Education News* [website], <https://twinkl.co.uk/news/school-trialling-insects-on-the-menu-to-support-sustainability-but-is-this-the-way-forward>, accessed 29 August 2024.

17 —'Wakes Labour Government blows over £400,000, on bugs for free school meals', 19 May 2023, *AberdareOnline* [website], <https://aberdareonline.co.uk/2023/05/19/wakes-labour-government-blows-over-400000-on-bugs-for-free-school-meals>, accessed 29 August 2024.

have the enzymes and hormones to process newfangled stuff. Surgeon Captain Cleave, a medical specialist in the Royal Navy based in Hong Kong, was the first to warn of the dangers of sugar and refined carbohydrates in his book *The Saccharine Disease*. Some five years after their introduction in Hong Kong he noticed a rise in heart attacks, and on further investigation even more problems came to light.[18]

A mountain of evidence exists that ultra-processed foods are bad for us. Including them in school dinners and dropping the food precautionary principles means we don't care about children's welfare at all. It's swapping healthy food for unhealthy food. Can't any teachers see that?

A long time ago I signed a petition not to boil lobsters alive. Boiling up grubs is no better. On the other side of my garden there's a park. Dragonflies, mayflies and butterflies fly round the pond. Taking these beautiful creatures and boiling them in urns makes me feel sick. And the energy required, boiling them, grinding them to powder, would be enormous.

Duncan Farrington has got rid of all carbon emissions from his farm.[19] If they really believed the 0.04% CO_2 in the atmosphere was a threat to us they could have asked other farms to follow suit. So why this fixation on destroying them? It doesn't make sense. Why bludgeon it as the only way? Why this unhealthy obsession with carbon and that it's bad for us? Plants love it, and it has nothing to do with global warming. Carbon dioxide levels have changed throughout history and there's no correlation with climate change. It's the sun that

---

18 —T. L. Cleave, 'The Saccharine Disease', *McCarrison Society* [website], <https://mccarrison.com/free-libraries-2/mccarrison-library/the-saccharine-disease-t-l-cleave-m-r-c-p-lond>, accessed 29 August 2024.

19 —Professor Andrew Barnes, Professor Bob Rees, Professor Mads Fischer-Moller, Dr Adrian Williams, 'Zero carbon farm', *Royal Geographical Society* [website], <https://rgs.org/about-us/our-work/sustainability/39-ways-to-save-the-planet/zero-carbon-farm>, accessed 28 August 2024.

makes us hot or cold, stupid.[20]

Cutting down on carbon means avoiding fossil fuels and turning to renewables instead. There are over 11,000 wind turbines in the UK. At times they generate too much energy for the National Grid to handle so they have to be 'curtailed', (government-speak for deactivated). The government stumps up the money out of our taxes. Research found that from January 2021 to April 2023, £1.5 billion was spent to curtail more than 6.5 TWh of wind power leading to 2.5m tonnes of emissions that could have been avoided. As more turbines are built the costs across the Scotland–England boundary are set to increase substantially. They could surpass £3.5 billion in 2030 and this could result in a nearly £200 increase in annual electricity bills for British households.[21]

Sorry to bore you with all this, but you should also know that each wind turbine uses around 187 gallons of grease, 40 gallons of hydraulic oil, 106 gallons of gear oil, 1585 gallons of

20 —'Climate the Movie—Renowned Scientists Speak Out Against the Climate Change Scam', 23 March 2024, *The White Rose* [website], <https://thewhiterose.uk/climate-the-movie-renowned-scientists-speak-out-against-the-climate-change-scam>, accessed 29 August 2024.
—*Clintel (Climate Intelligence)* [website], <https://clintel.org>, all accessed 29 August 2024:
—Guus Berkhout, 'There is No Climate Emergency, a Message to the People', 22 October 2022, <https://clintel.org/there-is-no-climate-emergency-a-message-to-the-people>;
—Clintel Foundation, 'All things Equal', 4 June 2024, <https://clintel.org/all-things-equal>;
—Clintel Foundation, 'Carbon dioxide and a warming climate are not problems', 31 May 2024, <https://clintel.org/carbon-dioxide-and-a-warming-climate-are-not-problems>;
—Andy May, 'Facebook Censorship of "Climate: The Movie" due to a Science Feedback "Fact Check"', (Facebook's censorship is totally out of hand...), 15 April 2024, <https://clintel.org/facebook-censorship-of-climate-the-movie-due-to-a-science-feedback-fact-check>;
—David Siegel and Prof. David Dilley, 'No Correlation and No Causation Between CO2 and Temperatures', 12 August 2024, *The White Rose* [website], <https://thewhiterose.uk/no-correlation-and-no-causation-between-co2-and-temperatures>, accessed 20 September 2024.

dielectric fluid whatever that is, 793 gallons of diesel fuel, 243 pounds of sulphur hexafluoride, 357 gallons of propylene glycol, 48 gallons of ethylene glycol, and they only last about 25 years.

After all they've said about getting rid of fossil fuels!

Moving on to chemtrails, you must have seen them, the white streaks in the sky behind planes.[22] [23] The normal airlines don't do them. I'm talking about clandestine aircraft. They

21 —Janet Richardson, 'Wind Turbine Information: A Guide for the UK in 2024', 21 June 2024, *The Renewable Energy Hub* [website], <https://www.renewableenergyhub.co.uk>, accessed 29 August 2024.
—Lorenzo Sani, 'Gone with the wind?', 15 June 2023, *Carbon Tracker Initiative* [website], <https://carbontracker.org/reports/gone-with-the-wind/>, accessed 29 August 2024.
—Matt Mace, 'Report: Wasted wind power costing Britain £1.5bn', *edie* [website], <https://www.edie.net/report-wasted-wind-power-costing-britain-1-5bn>, accessed 29 August 2024.
—Emma Gatten, 'Wind Constraint', 15 June 2023, *electricity info* [website], <https://electricityinfo.org/news/wind-constraint>, accessed 29 August 2024.
—'Images of the Week—Climate Change?', 5 June 2024, *The White Rose* [website], <https://thewhiterose.uk/images-of-the-week-climate-change>, accessed 29 August 2024.
23 *The White Rose* [website], <thewhiterose.uk>, all accessed 28 August 2024.
—'Chem Trails Are Blocking the Sun', 13 May 2022, <https://thewhiterose.uk/chem-trails-are-blocking-the-sun>;
—'How Does the Sky Look Today?', 01 May 2023, <https://thewhiterose.uk/how-is-the-sky-looking-today>;
—Dr Vernon Coleman, 'Have You Ever Wondered Why the Sky Is Full of Chemtrails?', 11 July 2023, <https://thewhiterose.uk/have-you-ever-wondered-why-the-sky-is-full-of-chemtrails>;
—'Don't Look Up!—Airline Pilot Exposes Truth about Chemtrails', 26 August 2023, <https://thewhiterose.uk/dont-look-up-airline-pilot-exposes-truth-about-chemtrails>;
—'Irrefutable Film Footage Of Climate Engineering Aerosol Spraying', 5 February 2024, <https://thewhiterose.uk/irrefutable-film-footage-of-climate-engineering-aerosol-spraying>;
—'Operation Indigo Skyfold—Chemtrails Must Be Stopped', 3 April 2024, <https://thewhiterose.uk/operation-indigo-skyfold-chemtrails-must-be-stopped>;
—'The Truth About Chemtrails From Pilots', 13 June 2024, <https://thewhiterose.uk/the-truth-about-chemtrails-from-pilots>;
—'I Spy With My Little Eye: Chemtrails Everywhere—18 July 2024,

seed the clouds with salt to reflect sunlight upwards instead of down here. So we don't overheat. Nobody's been asked if they want it and it's supposed to be secret, but we can all see them, we all know it's going on.

'No such thing,' said my MP when I complained to him about it. I wondered if he was myopic, but as he goes to cricket matches he's obviously able to see balls flying about.

There are several problems with this approach. Blocking out sunlight means people get less vitamin D which we need to make our immune systems work. When we don't have enough we catch colds and flu or even worse. Plants also need sunlight to be healthy.

And blocking out the sun makes life miserable. It's going on all over the world. A lady from East Coast Australia complained on the internet that the six months of sunshine they enjoyed every year had been replaced by dreary grey clouds. Dreary grey clouds and chemtrails. She included a picture to show us.

Cloud seeding can go wrong. It's caused floods in Dubai. Floods have been ruining farmers' crops in this country too. How can we tell whether it was a natural event or geo-engineering? There's so much censorship going on we've got to the point where it's difficult to know the truth about anything any more.

Planes spray other chemicals too, besides salt. Spanish crops were affected when farmers found higher levels of aluminium in the soil. It was even in the soil on organic farms that had had none before. Belgian farmers are complaining that chemtrails are making their cattle sick, and one of my friends said he used to have aphids in his garden but not any more.

What good does it do dropping chemicals on us? What

<https://thewhiterose.uk/i-spy-with-my-little-eye-chemtrails-everywhere>.

has it got to do with climate change? What has it got to do with anything? It's poisoning the air and the plants in our gardens. My cat likes to sit there and horrors, sometimes she eats the grass!

End of discussion. Finis. That's it. There's no way I'd support anything that hurt my cat. This is madness. Can't people see that? Utter madness.

Let's all stand at our front doors and clap for farmers, clap for cows, clap for all the animals and plants that make our planet so beautiful.

And you can put that in your pipe and smoke it, Mr Monbiot.

The polar bears never took the fur coats off to do a bit of sunbathing and are still freezing. The ice and the polar bears are still there.

—*Roy M McIntosh*

XXVII.

Hot potato

By Roy M McIntosh

Yes, the *Green Cheat*. Where and when does one start with this 'hot potato' as there are so many liars involved and people with vested interests and hidden agendas, no doubt. When did it start? Maybe with the odd smart loony caveman telling the others no, we cannot have a fire every day as we shall burn all the trees! That never happened, so the caveman was a conman and that was way back, so nothing new regarding cons and scams.

Now in my short time I have seen and heard what I take to be satanic lies being told for the benefit of the green cheats. Get a diesel car, as is better for the planet; the sea is going to flood parts of London and New York and islands shall be under water! I have asked in the past where is the water going to come from for the sea level to rise so high? The area covered with ice is minuscule compared to the area of oceans and seas, so in my own brain I thought that would only make the sea rise a couple of inches at most, if every bit of ice melted. There shall be no ice left and no polar bears... Well, the polar bears never took the fur coats off to do a bit of sunbathing and are still freezing. The ice and the polar bears are still there.

No doubt earthquakes and earth movement shall have caused changes at some places. Erosion on coastlines also shall have changed areas in small ways.

Just last week I jumped on the train at Edinburgh to go to Inverness and took my bicycle and cycled to John O'Groats from Inverness, boy it nearly killed me as I found out the true meaning of hills. When I got to John O'Groats, it took me three days of green energy—my own! I decided to carry onto Thurso to get the train back to Edinburgh. I had a couple of days in Thurso. I do not have a phone and did not have a map and travelled on luck and instinct. In a hotel I was in, I looked at a map of the area and saw that Scrabster was only a mile or so away. Back in the days of my working life I had repaired fishing boat engines and I heard the fishermen say they had looked into Scrabster. That gave me the prompt to go and see the place as it was just 10 minutes along the road.

I am glad I took the short journey along to Scrabster, as when there I found out a couple of things that I reckon shall fit in *The Green Cheat*.

For many years I have always known that 'wee porkies' were being told. Like get a diesel car, as it is better for the environment than your petrol banger and that was in the 1980s. A contact of mine worked at a council or government owned old cathedral and they were told to separate all the rubbish that had to go out. Plastic, paper, glass, etc. and they were doing that, but one day by chance, he saw the rubbish getting lifted and it was all going into the back of a truck. He asked why this was happening and was told, oh, it all goes to landfill.

TRIP TO SCRABSTER

On the Thursday, 4th of July 2024, I popped along to Scrabster on my bicycle as was only a 10-minute cycle, and although weary from previous days cycling, I knew I could manage that. Got there as the Orkney ferry was firing up and getting ready to depart.

Once down in Scrabster I saw it was just more or less one street with the harbour. Think there was a pub (The Ferry Inn) as they go with harbours. First couple of things I noticed was a big fishing boat tied up and not far from it a mountain of what I thought were pit props for the coal mines, but my mind then said there are no mines left? Maybe they export them? I cycled around a bit and after a while got speaking to a couple of locals and one was working, shore side fish worker he is, but he took time to speak to me. I asked them about the big fishing boat and the reply was not kind words. Turned out to be a Spanish owned fishing boat and between the fellas they told me that the crew are all from over seas and that they contribute nothing to the community. According to them a truck brings all the boat supplies and takes away the catch. I would imagine that is back and forward to Spain. All sounds like more green shenanigans. The fella working departed with his last words: since BREXIT the foreign boats have more rights and freedoms in the Scottish waters than our own boats. I took the words in good honest faith.

Other fella and I chatted for a bit and he told me he lived just up the hill above the town. Was a top part to the town with a couple of streets but never wandered there as was more interested in the harbour and boats. We chatted about this and that and suddenly a thought came into my head. Whilst I had been cycling from Inverness I had seen big trucks here and there loaded up with what I thought looked like pit props. Then the sudden thought in my head. I just said, see the mountain of wood, do not tell me it is for biomass. He smiled and said yes and this has been going on for ten years. He went on to say forests had been flattened in Caithness and he named one large forest at Berriedale on the A9 south of Wick. Next day I spoke to a local man and I asked him about the chopping down of the forests and he confirmed the Berriedale one as he drives that way and he said before you saw trees, now it is bare land looking over to the mountains. Before I left the man in Scrabster on the Thursday I mentioned Dounrey and the contaminated beach. He said that is another story. I have had a read up on the cutting down of trees and one of the excuses mentioned is to reduce carbon and other greenhouse gases from escaping. Sounds like more fuel for the green cheat.

On Friday the 5th of July, it was a cold morning with showers but rather than be cooked up in a hotel, I still went out. Another look around the town and went into local museum that I had poked my nose in on the Thursday and said I would be back. Stuck to my word. Had all the usual stuff about local celebrities and standing stones, etc. Also a large area about the Dounrey Nuclear Plant which is getting decommissioned.

When I left the fella at Scrabster he did say that Dounrey was another story. He did say that there were people against the shipping of nuclear waste by ship from Scrabster, as they had to bring it to harbour by trucks and people were worried. Allegedly they said they would stop moving the waste but with him living there he said that he reckoned they were doing it under the cover of darkness.

On the Friday I went down by the river and had met a fella and we had spoken about the biomass trees. I had heard in the past that there had been an explosion at Dounrey, he reckoned more than one, but all kept hush-hush.

As said, the museum had a big area about Dounrey but the thing that caught my eye was the photograph of after the explosion and the other of the contaminated beach.

In the past I have read that nuclear was going to be so green and also so cheap that it would not be worth metering it. I do not recall free electricity!? As for green, that claim was maybe a bit hasty and maybe poor management caused the explosion and contaminated beach. I have read that there have been concerns about nuclear contamination in the Irish Sea from Sellafield and that has been for decades. Can it get into the food chain? I am not sure if there are fishing grounds near Scrabster, but I am sure I saw lobster or crab creels at Thurso and Scrabster, and one can only hope that there is not fishing too close to Dounrey or Sandside Beach which is the contaminated beach. But how much contamination is there and is it not further afield? Also, how long shall it take for it to clear up? I have also heard there have been cover ups about this so it all sounds very fishy. Hope the fish are OK?

On the Saturday I was at the station to get my train. Cancelled. Grrrr! But it gave me a last minute chance to ask a few questions. At the station was a local lady and I asked her if Dounrey had been good for Thurso. She said for a few and it had increased the population but also changed the place. I never asked if for better or worse. She did throw in that she shall be glad when it is gone, but also said that how long shall the contamination last? Also said that there was a high level of cancer.

Now I am no expert on the ins and outs of how the world worked and never went to university, but through out my life worked and had friends who were a lot smarter than your average university lecturer or politician and they worked for their living and did not scrounge off the public purse. What I found was that they did not waste material as it cost money and in general were much more friendly to Old Mother Earth. Saying that I saw things that were wrong and in the past have written about them. Here is one of my writings.

The world is a lovely place and we know of the dangers in the oceans and jungles but the real danger is the people!

Old Mother Earth has been battling the poisoning and

scarring that has been going on since day dot! In the last hundred years that battle between people and Old Mother Earth has intensified, who knows how many times. I have seen bits of it as I lived not far from a coal mining area with the slag bings reaching down to the sea. Worked in an open cast iron ore mine in Australia where blasting and digging showed no respect for man nor beast and destroyed historical sites, no doubt. I was there mid 70s and I was told they expected mining to last around 25 years. It is still going and what year is this!? Then I have worked on ships where bilges were pumped into the sea and we dumped all sorts of rubbish over the side. Also involved in the fishing industry and saw fish being landed but as was too many dye poured over them and they were taken and dumped. What damage the military has done by wars and also dumping at sea can be anyone's guess! The big companies have added a lot to the pollution and it shall be clear that the oil and pharma companies shall be at top of the list! I did work at a pharma company for a short time and they used to dump all sorts of rubbish into the sea when the tide was going out.

Now I have read that some countries, the likes of Holland and Ireland are blaming the cows for creating pollution from passing wind. I started looking into that and wow, I found out that the Americans slaughtered 50 million buffalo in the 1800s. If I am right with that, why did the 50 million buffalo not destroy the earth? I am sure they would have passed wind.

Time shall tell us if people shall sink or swim, but for sure, I reckon Old Mother Earth shall still be floating along.

The world is full of great criminals with enormous power, and they are in a death struggle with each other. It is a huge gang battle, using well-meaning lawyers and policemen and clergymen as their front, controlling papers, means of communication, and enrolling everybody in their armies.
—*Trappist Monk, Thomas Merton, 1962*

XXVIII.

Step by step

By Eileen Coyne

'If one is free at heart, no man-made chains can bind one to servitude.'—Steven Biko

I remember as a relatively young child, at the age of ten or eleven, struggling to read through a three volume, hardback collection of the Lives of the Saints. It was hard going, a vast collection of men and women, many being killed in strange and agonising ways, all steadfast in their love of God, their eyes fixed towards heaven while they experienced hell on earth. I admired them but was not keen to emulate them; maybe I should have stuck with something gentler like C S Lewis. But that personal relationship with God had formed early with me and I was eager to know what God wanted of me; what my purpose was. Now, I know.

I was a teenager in the early 1980s, a wonderful time. We had LiveAid; UB40's 'I in 10', The Specials 'Ghost Town' and the wild energy of the late Punk era mixed with social activism and a collective fervour that we could effect change. Music is what kept me sane in my turbulent teenage years and there were powerful, positive, lyrics which stayed with me. Even if I haven't heard a song from that period in 40 years, as soon as it starts to play, the lyrics come back to me word for word, along with a rush of energy and the faint scent of my youth plays in my nostrils for a time.

Many of the films too had positive messaging; films like Star Wars: 'Use the force, Luke' but they could be deceptive. The baddies looked evil and the goodies always won in the end. Real life had different lessons to teach me.

As I hit my twenties, I came to realise that, as people reached what I would call a certain level of spiritual maturity, they would normally be taken out of the equation, or killed, to put it bluntly. There are numerous examples. Jesus, of course, but Gandhi (assassinated), Martin Luther King (assassinated), Bob Marley (assassinated), Oscar Romero (assassinated while celebrating Mass). The list is truly frightening and these are only the highlights; the people well known. How many hundreds or thousands of others met similar fates; unknown, unremarked but remarkable souls.

Others seemed to meet with strange accidents. The Trappist Monk, Thomas Merton, wrote in a letter in November 1962: 'The world is full of great criminals with enormous power, and they are in a death struggle with each other. It is a huge gang battle, using well-meaning lawyers and policemen and clergymen as their front, controlling papers, means of communication, and enrolling everybody in their armies.' Thomas Merton died in questionable circumstances in 1968. The inspirational Steve Biko (who I have quoted above) died whilst in police custody in South Africa in 1977. The message was clear: 'If you raise your head above the parapet, don't expect to last long.'

So how did these realisations affect me? On the one hand, I had a good career in London as a Shorthand/PA, just an average commuter and office worker. Spiritually, I realised that the world was a dangerous place to be a good person in and by 'good' I mean a spiritual activist, trying to hold up a mirror to the world, to defend human rights and human dignity and to make a difference, in however small a way. I realised that good people tended to rest, content, within themselves whilst people who had chosen a dark path were never still, never satisfied, always hungry. I determined to stick by my principles and tend my garden, the small space I found around me; to try and support and empower people and stand up to injustice. It didn't cost me much, to be honest. Yes, once or twice I refused to join in the office politics and stood by the latest person being isolated and victimised, knowing that I would be next on the list. But I was very good at my job and jobs were plentiful in London so my occasional moral stances were mere short-term inconveniences, truth be told.

I suppose I would describe myself as half-awake, or sleeping with one eye open, until the events of September 2001, the Twin Towers in New York. I didn't need to be an engineering expert to know what I was witnessing were buildings in free fall due to controlled demolition and knowing that it would take months to set the necessary charges to enable buildings of that size to collapse within their own footprint. It hit me hard as I had been an office worker. I could visualise being there. First cup of coffee in hand, planned work for the day at my side but scanning emails to see if any emergencies had cropped up which would disrupt my proposed work schedule. The terror those poor, trapped people must have felt; the awful decisions many would have to make; even the date 9/11. 911 is the emergency number Americans dial in time of need, like 999 in the UK. I sensed true evil then, something now only half masked, leering, engaged in gleeful destruction and confident enough to begin to emerge from the shadows.

My response to this was to try and remain more awake, to question more, to shift my scant charitable resources to coalface organisations rather than those who invested in high paid CEOs and plush offices. I realised that there were huge issues with the education system; it was a form of parrot learning slanted towards those with retentive memories, with little being taught in the way of ethics, morals or practical skills. I realised that the medical professionals had largely become drug peddlers. There was little compassion within the structure of society for those with mental illness. Funding had been cut to almost nothing and the solution seemed to be to wait until men and women broke the law before carting them off into an even more traumatic environment; prison. Every attempt was being made to nullify religion. The extended family (a vital support network) had broken down and then the nuclear family had largely been broken down too. My Father had been a labourer but had still managed to get a mortgage on a house in London on his wage. Now, those nuclear families that still had two parents were in a situation where both were working full time to make ends meet and paying someone else to mind the kids. Consumerism had gone totally crazy. The messaging had turned negative: 'You are not fit enough; you are not pretty enough; you are not rich enough; you are not enough'.

Spiritually, I had found the mystics. I especially resonated with John of the Cross and embraced his 'todo y nada' philosophy. I realised that we were, as individuals, both totally insignificant and yet vitally important on the world stage. We were insignificant as one of billions of humans, yet, each one of us was created to a unique blueprint by a loving Creator, we were relatively powerless and weak in ourselves yet all of us had a divine connection to God and could be a conduit for his Grace. We were all 'todo y nada', 'everything and nothing' at the same time. I had been granted a profound experience; an inflowing of the unconditional love of God that cured me of many fears and all doubt. Unfortunately, I didn't grow angel wings and, day to day, my greatest battle was still with myself (or should I say my false self, the 'ego') but it was something I never forgot.

So the Covid-19 years and all the manipulations and lies that followed on many topics did not shake me as much as it must have shaken many people, but there can be no doubt that it has had its impact on everyone. I feel that the very real evil in this world is finally totally unmasked and is hell-bent on our wholesale destruction. However, I would not underestimate the power of prayer to counter this evil. In a busy life, if you spend even five minutes a day sending out a prayer for love and healing for the world, like the butterfly's wings, you will create a tsunami of love somewhere on the planet (and it will also help you to centre yourself). All the institutions on which people formerly placed their trust have shed their quasi-benevolent skins and we are in the crucible. There can be no doubt about that. I meet this threat to my existence as a post-menopausal, overweight woman in her late fifties, a couch potato if ever you saw one, totally unremarkable except for the fact I'm pretty good with the written word. I meet this threat as a spiritual being, loved by God and dedicated to standing up and resisting every attempt to demean my autonomy as a human being. My head is firmly above the parapet and I don't expect to survive but, let's face it, I'm on the downward slope anyway. I never thought I would live to see such times, but there is a part of me which recognises that, by my life experiences, I have been led, step by step, to this very spot. I will meet every challenge head on, not in any dramatic way but with prayer, foresight and kindness. If I can show one or two young people that there are alternative paths to take or ways to think, if I can feed one or two people and help them through the manufactured famine that is rapidly unfolding, if I can heal one wound or give one moment of comfort to someone, it will be enough; enough for me and I have confidence it will be enough for the unconditionally loving God I will return to.

The dramatic pictures of melting ice and calving icebergs and polar bears swimming between melting ice floes give the impression that the ice is constantly melting.

—*Keith Brown*

XXIX.

The green apocalypse

By Keith Brown

Keith has taught History and Economics for over 30 years; a degree of historical perspective being, of course, a useful tool in debunking a variety of modern nonsense. Keith is also an author of the children's book; *A Bigger Bird!*

THE CLIMATE LIE IS DEMONSTRABLY UNTRUE

LIE 1—'INDUSTRIAL CO_2, AS A DOMINANT PART OF THE EARTH'S ATMOSPHERE, CAUSES THE CLIMATE TO CHANGE'

The concept that there is some huge amount of CO_2 in our atmosphere is simply untrue. 0.04% of the earth's atmosphere is CO_2. There has been an increase in the amount of CO_2 in the atmosphere but this tiny proportion is unlikely to be responsible for any adverse effect; CO_2 is, of course, crucial for life on earth as it is required by plants for photosynthesis. The climate has always changed, ice ages and warmer periods alternate. We are currently coming out of the mini ice age of the sixteenth and seventeenth century, i.e. the climate started warming before industrialisation. The Roman warm period and the medieval warm period both had higher average temperatures than today and in the prehistoric period there was a period when the Antarctic had no ice at all due to much higher temperatures than today as evidenced by the fossil record. In the 1970s there was a scare about global cooling which experts predicted would lead to falling sea levels. This was based on a fall in average temperatures between 1920 and 1970. In other words the earth's climate has always changed, there is no logical link or even long-term correlation between CO_2 levels and climate. The idea that more CO_2 causes higher temperatures is simply unprovable as the earth's atmosphere cannot be replicated in a laboratory. There is not even a long-term correlation; what do the experts base their view on? Computer modelling; which if you feed certain data in you will get a certain result; this is the equivalent of looking at a purple coloured house with a white dot on it and by focussing purely on the dot claiming the house is white.

LIE 2—'SEA LEVELS ARE RISING'

This is against the evidence of people's own eyes. Walk down to a beach and observe; the sea level is exactly where it was 50 years ago and more. There are places where the land is sinking—the eastern seaboard of the USA for example, which gives the appearance of rising tides; there are other places where the land is rising, e.g. parts of Sweden, giving the appearance of falling sea levels. Another example is Harlech Castle in North Wales which was built much closer to the sea than it is today. The sea level is simply not rising.

LIE 3—'THE ARCTIC AND ANTARCTIC ARE MELTING'

A version of this was peddled by the then Prince Charles at the EU parliament in 2007 where he stated that all Arctic ice was liable to have melted within seven years! Arctic ice does indeed melt, in the summer. It also, of course, builds back up in the winter. Antarctic ice melts in our winter and, also, of course, builds up again in our summer. The dramatic pictures of melting ice and calving icebergs and polar bears swimming between melting ice floes give the impression that the ice is constantly melting. We are shown pictures of Arctic ice melt in, for example, August, while also being shown pictures of Antarctic ice melt in, for example, December. Were we to be shown Arctic ice growing in, say December, and Antarctic ice growing in, say, August, our impressions would be entirely different.

LIE 4—'THE UK IS PARTICULARLY CULPABLE FOR CLIMATE CHANGE'

This, of course, assumes that the climate is changing due to industrial CO_2. It also considers being responsible for industrialisation, inventions and modern conveniences to be a purely bad thing. It is also untrue even on its own terms as the UK emits about 1% of current industrial CO_2 and China, for example, has built over 100 coal fired power stations in the last 15 years while we commit economic suicide over an unprovable theory. China has emitted more CO_2 in the last 20 years than the UK has in its entire history.

LIE 5—'REFUSING FRESH OIL AND GAS LICENCES FOR THE NORTH SEA OIL AND GAS FIELDS IS ENVIRONMENTALLY FRIENDLY'

If all oil and gas exploration stopped worldwide (which it won't) and if CO_2 emissions actually caused climate change (for which there is no evidence that they do) then this might hold water. However, refusing North Sea oil and gas licences leads to the import of more oil and gas, bad for the balance of payments, bad for UK employment, bad for UK living standards and bad for CO_2 emissions (not that that is important anyway but it's what it is done in the name of).

Lie 6—'Green energy is good for the economy'

Switching from systems that work, i.e. gas-fired power stations to systems that don't, i.e. wind which only works, well, when the wind blows and solar which works, well, when the sun shines; raises energy costs, discourages firms locating and producing in the UK and makes us all poorer. Electric cars are heavier than petrol and so require more energy to move them. Electric cars are more likely to catch fire than petrol. Electric cars require lithium which causes environmental damage when mined. Electric cars are often imported which is bad for the balance of payments, UK employment and is heavier on resources than domestic production. Electric cars run, naturally, on electricity of which one third is generated by gas-fired power stations anyway. They are also unaffordable for a proportion of the population.

Lie 7—'The UK is seeing record temperatures and extreme weather events'

To be valid, temperature comparisons must be like for like. Urban readings are higher than rural. Some parts of the country have higher readings than others; due to the Gulf Stream the West, on average, is warmer than the East. Thus 'record' temperatures can easily be announced with monotonous regularity either by taking a set of data for which the data only goes back a few years or by comparing readings which are intrinsically higher due to where and when they are taken. The number of wildfires has actually fallen in the last twenty years and one has only to think back to the hot dry summer of 1976 or the Great Storm of 1987 or to look in the historical data at the wet summers of the mid-Tudor period or the Great Storm of 1703 or the east coast storm surge floods of 1953, to see that current UK weather is in no way shape or form peculiarly extreme at the moment.

IF IT'S ALL LIES WHAT IS IT ALL ABOUT THEN?

There is a well-worn phenomenon called millenarianism. The Christian concept of the apocalypse and original sin is deep in our psyche. In an increasingly secular age, the global boiling—yes that phrase has actually been used—idea is a faux perversion of the Christian religion. We are guilty, so the thinking goes, of the original sin of industrialisation, we can repent and make amends by driving electric cars, not flying, covering the countryside in windmills and generally making ourselves poorer. If we do not atone, we will face an apocalypse where people violently compete for food, energy and water and poke each other in the eye with sticks! (Just Stop Oil and Extinction Rebellion promote exactly this idea).

In the year 1000 the world was projected to end, in the 1840s people got out of the train at Boxhill Tunnel and rode round to the other end to board the same train as experts had told them that travelling at speed through a tunnel would lead to death, in the year 2000 all computers were going to fail due to a computer error, in 2020 500,000 people were going to die of Covid-19—the official figure was 229,000—who actually died within 28 days of a positive Covid test, not of Covid-19.

The climate lie has become a vested interest. If you own wind turbines, import electric cars, wish to damage liberal democracy or fit heat pumps, then naturally you will, if immoral, lobby for more 'climate action'. Someone who is a 'climate scientist' needs there to be a climate emergency to stay in a job.

HOW TO CALL OUT THE CLIMATE LIE

Once people are aware that CO_2 is only 0.04% of the atmosphere, once people are aware that sea levels have not risen, once people are aware that the Arctic has not melted, once people are aware that the UK generates only 1% of worldwide CO_2 emissions anyway, once people are aware that rational, sane, people do not automatically believe the climate lie; then the increasingly hysterical projections of the climate activists, funded by Green billionaires such as Vince Dale, will become seen for the absurdities that they are. If we, the many, do not accept the climate lie, if we the many show we do not believe in the climate lie, if we the many refuse to buy an imported electric car or swap our efficient and effective gas boilers for dodgy heat pumps, then the climate lie will fall apart.

It was once widely believed that the sun revolved around the earth and those who questioned this concept were persecuted. This seems absurd today; in the same way, the climate lie will seem absurd to the future and the sooner people go down to the sea, look at the beach and see that the sea level is not rising and the earth if not boiling, the better for us all.

They regard the 'common people', those people in the world who are not billionaires or trillionaires like themselves, with considerable distaste, and wish that there were far fewer of us around.

—*Ralph Prothero*

XXX.

The religion of the rich

By Ralph Prothero

Ralph Prothero is not now, nor has ever been, a member of any political party.

Usually they are small islands with few people, in oceans far from the mainland, like the Cocos Islands in the Indian Ocean or the British Virgin Islands in the Caribbean. They are the ideal home for tax havens, the havens from high levels of tax, where global plutocrats store their wealth.

With their insular government making an independent tax regime possible, these islands charge the global moguls, who own the world, tiny rates of tax or no tax whatsoever, in return for the moguls' help in making life comfortable for the island people.

The moguls' generosity ensures the islanders' fierce loyalty, any islander who turns whistleblower or informer against this hidden world can expect to be shunned and ostracised, or worse.

But occasionally we do get a glimpse behind the curtain. One such glimpse was through the release of the Panama Papers onto the internet, when files from the Panamanian offshore lawyers' firm of Mossack Fonseca were made public. The files showed not just tax avoidance, but tax evasion, by the global super-rich. In the immortal words of Leona Helmsley, the hotel heiress and one-time Queen of Mean on American TV: 'Only the little people pay taxes.'

When the wealth of the billionaires and trillionaires is hoarded on the mainland in hedge funds (the biggest hedge fund is BlackRock in the USA) ownership is hidden behind anonymous shareholdings in faceless companies whose interlocking ownerships make a Rubik's cube seem easy to unravel.

Thus the global moguls, the international plutocracy, safeguard their ownership of the world, especially their ownership of the mass media and their control of its content, and their control over what is known as 'the news'. Hence the incessant Green propaganda we are bombarded with day in and day out.

They easily buy politicians as well (politicians need millions for their election campaigns), apart from those politicians too rich to be bought, like Donald Trump, who hence earn their undying hatred.

Naturally the plutocrats flock together as birds of a feather at their exclusive waterholes. One such venue is at Davos in Switzerland for the annual congress of the World Economic Forum, where the minimum entrance fee is fifty thousand dollars. And naturally they regard the 'common people', those people in the world who are not billionaires or trillionaires like themselves, with considerable distaste, and wish that there were far fewer of us around.

People believe what they want to believe, and as the philosopher David Hume wrote, 'Reason is the tool of the Passions', so the globalist Green plutocrats believe in global 'overpopulation', because they loathe the 'common people' of the world, whom they see as a 'swinish multitude'.

That is why the global moguls dislike nation states with governments elected by votes cast by the 'common people'. The plutocrats much prefer supra-national organisations like the United Nations or the World Health Organisation which are untainted by the vulgarity of democracy. And it is through supra-national organisations that they expect to achieve the globalist dream of a global world government.

A growing global government already exists in the United Nations Organisation. The ideology of this global UN government is Greenery, the rigid dogma of Green pseudo-science, with its crazy claim that human beings control the climate of Planet Earth.

When Green politicians say that the climate science is 'settled' they show their ignorance of what science is. It is religion that is always settled by unquestioning faith, whereas science is always unsettled by its critical questioning of established dogma and sceptical attitude towards it. The blind belief in Green pseudo-science displays the triumph of dogma over science in the modern world.

The major theme of Greenery is its claim that carbon dioxide ('Carbon' in Greenspeak) is threatening life on earth with extinction (hence the disruptive stunts of Extinction Rebellion). The purpose of this vilification of carbon dioxide is the destruction of reliable energy supplies for the Western world. Coal, oil and gas ('Fossil Fuels' in Greenspeak) are reliable sources of energy, Green windmills and solar panels are not. The Green Net Zero policy that bans fossil fuels applies only in the West, while China builds coal-fired power stations with reckless abandon.

Condemnation of carbon dioxide leads to condemnation of Fossil Fuels which produce carbon dioxide when they are burnt. Net Zero's ban on Fossil Fuels will ruin the West by making energy supplies much more unreliable and expensive, which is the purpose of Net Zero. Even though nuclear reactors don't produce carbon dioxide, Greens in the West campaign against nuclear power, because it is a reliable source of electricity.

Using the preposterous pseudo-science of the UN's Inter-governmental Panel on Climate Change ('global boiling') the global moguls intend to wreck the West. That is because the votes of the 'common people' in the West are seen as the chief obstacle in the way of global government and its enslavement of the 'common people' of our world. Elsewhere than in the West the 'common people' do not have an effective voice, so the West is the main target for Net Zero and the other stratagems of the globalist plutocrats.

Other stunts that the green global moguls have staged include the Covid Terror of 2020, the purpose of which was to get messenger RNA (mRNA) jabs into the arms of the 'swinish multitude' of 'common people' worldwide.

The messenger RNA in the genome-altering jab artificially alters the jabbed person's DNA by a process called Reverse Transcription. These mRNA jabs were passed off as 'Covid vaccines', that would save us in the years of the supposed Covid-19 pandemic, when all-cause total death rates rose little, if at all.

Once people's DNA has been artificially altered by the jab they become 'human organisms' in Human Rights law, and lose the human rights that they had before as 'human beings'. Their children will inherit their artificially altered DNA and so will be 'human organisms' also without human rights. Thus atrocities committed against them in the future by the global green government won't be illegal, as they won't have any human rights.

Once cash is abolished and all money becomes Central Bank Digital Currency which the 'common people' get through their mobile phones, a tap on a computer key at a central bank database will stop the 'common people' from getting any more money.

The consequent death from starvation of billions of people is what Greenery, the Green religion, has always wanted. The Green dogma of 'unsustainable over-population' shows the loathing of the Greens for the 'swinish multitude' of human beings on Planet Earth.

Back in the 1970s the Ecology Party in Britain, forerunner of the Green Party today, said that the ideal human population of Planet Earth should be three hundred million people. As there are now over eight billion people in the world there are more than seven billion people to get rid to get down to the ideal Green level of three hundred million people for a 'sustainable' human population on Planet Earth.

Greenery, the Green religion, is the religion of the rich, it is the delusion and deception at the core of the plutocrats' plot against the rest of humanity. It is proof of the infinite capacity for self-deception that is innate in human beings.

For the last few days the internet has been buzzing with people asking why has June this year been so cold when they tell us in reality we are in a heatwave. People have been saying instead of basking in the sun in parks they have been at home with jumpers and coats on in the home, others putting the heating back on and some even using hot water bottles. Yes, instead of lying on the bed with the covers thrown off gasping in the night time heat, I have been putting extra blankets on myself. All wrong for this time of year. It feels more like winter than summer. If, as they say, it's been hotter here this month than in Corfu, then it must be a lot colder in Corfu.

I assume Jim Dale from the MET office (not the comedian), constantly trying to do a good impression of him, will have the answers to make us laugh on these cold summer nights. It seems if in any one given month it's just one day warmer than average, by just a tiny amount, Jim and the BBC bark it's been the hottest for that month on record. They fail to say the other days for that month are average or even cooler it's that one day they seem to use to create their fear. Even National Newspapers are starting to say is it time to stop worrying about this non-existent climate crisis.

—*James McGuinn*

XXXI.

Weapons of control

By Ray Wilson

> Travelling around the world with his family as a young boy,
> Ray became fascinated by radio technology and electronics.
> This led him to pursue a career in engineering, where he tried
> to make a difference and in small ways challenge the regime
> that controls us all. When he's not working with electronics or
> writing stuff, you'll probably find him out motorcycling with
> his wife, walking the dog, or working on various projects. He
> finds solace in the simplicity of nature and the freedom of the
> open road.

In a dimly lit secret military bunker during the early 1960s, a group of high ranking officials sat around a table covered by maps, charts, and prototype electronic devices.

'What we are about to discuss will change warfare forever, but more than that, it will enable us to control enemy populations—or domestic populations if required,' said the general.

The admiral cleared his throat. 'I've read the preliminary report; it sounds a bit far-fetched, I thought.'

'Indeed,' said the good doctor, 'but imagine the ability to control the weather—to introduce changes greater than those we saw in 1945 using nuclear weapons—my opinion is that this technology is a game changer.'

The Major suddenly alerted, 'You're suggesting we could

control storms, hurricanes, that sort of thing?'

'Exactly,' the good doctor replied with a knowing smile. 'Imagine the strategic advantage that would give us in any conflict scenario.' The room fell silent as the implications of this new technology began to sink in.

'A severe storm or hurricane—yes, directed with precision towards a naval force could inflict more devastation than a conventional weapon,' the good doctor added.

'So we have the ability to change the direction of a destructive storm, guiding it to enemy concentrations,' the admiral tapped the map.

'Ground and amphibious operations revolutionised—fogs or clouds dissipated to clear the way for a surprise assault,' the general looked across at the map.

'Conversely, we could create solid overcast skies to conceal troop movements,' said the major.

'Perhaps magnetic and acoustic effects could be generated in such a way to cause an ocean wide sweep of mine fields—humidity ducts to modify the refractive index to influence radar and radio transmission,' the good doctor suggested.

'Imagine the advantage of manipulating weather patterns to gain a strategic edge in warfare,' the admiral added.

'This level of control over the environment could truly change the course of battles and conflicts,' the general concluded.

'The potential is immense; we must move forward with the utmost care and secrecy—this conversation never leaves this room.'

'Understood, general,' the admiral replied.

The coms phone rings over a crackly line.

'Admiral, sorry to interrupt—an urgent message from your wife—it reads—all well, it's a boy.'

The room falls silent as the officials turn their attention

to the various maps and charts.

The streaky mackerel skies of 2025 stretch over the farm and its surrounding fields. The year had started wet. Days of torrential rain, and fields were soon flooded. The floods were followed by days of blistering, intense heat that parched the land. Crops failed, and animals perished because of the extremes of the climate.

Two gentlemen are leaning on a metal gate.

'She called me out of the blue,' Dan explains, 'and asked if I was interested in putting the herd through a trial to stop my cows farting and belching, she said, emitting methane. 'Well, I am not sure,' I said. 'You would be amply rewarded', she went on to say. 'How do you monitor the results?' I asked her. 'Do I have to put a detector of some sort on the cow's tail? I am not sure they would like that.' 'No, no,' she said, trying to reassure me. 'So how are you going to monitor the emissions so you can see definitive results?' I asked.

'What did she say to that, then, Dan?' Reg asks.

'She explained that they would use satellite technology to monitor the emissions remotely,' Dan replies. 'It seemed like a pretty advanced system, but I still had my reservations.'

'I told her that I wasn't really interested, and she turned all mardy saying I would be forced to do it in the end, so I ended the call.'

'Mindless pen pushing officials,' Reg rubs his head.

'It's all about pushing vaccines—vaccines,' Reg continues, 'to prevent farting and burping in animals, a vaccine that targets methane-producing microbes, it's nonsense, for what?'

'Decarbonisation in global meat and dairy products, I suppose,' Dan replies.

'Yeah, we are next,' Reg says. 'I read something about some biotech companies working on human-targeted climate vaccines too. What's your take on that? Do you think this could lead to climate vaccine mandates?' Dan asks.

'That's a possibility. Especially with AZ calling climate change a public health crisis and the executives discussing vaccine innovation as a response to it. It makes me wonder if we might see climate vaccines becoming a requirement for travel, similar to what we saw with Covid-19 vaccine passports.' Reg looks concerned.

'Are you taking the devil's coin, Reg? That twenty acres of flat pasture is perfect for solar panels, isn't it?' Dan asks with a mischievous glint in his eye.

'I have been suffering with insomnia since I got the offer —easy money—but?'

'But what?' Dan asks.

'My dad gifted me the farm; he wasn't a farmer and brought the place on a whim when he went off the rails.' Sadness fills Reg's eyes. Dan's mischievous expression softens as he listens to Reg's story. 'Maybe turning it into a solar farm could be a way to honour your dad's memory and get an income for life,' he suggests gently.

'No.' Reg shakes his head. 'You don't understand; none of it makes sense—how much sunshine have we had this year?'

'Not much,' Dan admits.

'Exactly, look, I promised dad, oh, sod it, Dan. Dad told me on his deathbed that he had made a terrible mistake.'

'Go on,' Dan said, looking intrigued.

'You know he was a high ranking official?'

'Admirals Farm—I sort of guessed—hobby farmer living in town.'

'Dad told me they developed a system to control the weather; it worked brilliantly at first—it was used as a military weapon before it fell into the hands of the elites.'

'You are starting to sound like a conspiracy theorist now.' Dan laughs.

'Let's have a beer or two. Come over to the farmhouse. Forget what I said, Dan. I am just getting paranoid in my old age; it's worse since the wife died suddenly last year and the boys left.'

The clouds swirl around the farm, becoming darker and more ominous as they approach the building, adding to the mysterious atmosphere. A strange redness bleeds out from the cloud edges as the sun slowly sets. The wind picks up, carrying with it a faint metallic scent that sends a shiver down Dan's spine.

'I'll join you for a beer, but I can't shake this feeling that there's more to Admirals Farm than meets the eye,' Dan said with a curious glint in his eye.

I read an amazing story in the Telegraph today. After millions of years, scientists have just discovered hay fever is caused by climate change. Of course, it is and of course climate change exists. We see it every year, four times a year and have done since time began. Some people call it Spring, Summer, Autumn and Winter.

—*John Willis*

XXXII.

Climate change, what nonsense

By Harry Hopkins

> Harry is a retired furniture designer/ maker. Having passed
> through a successful career in industrial management, Harry
> now spends his time writing, gardening, enjoying the natural
> world and with like-minded friends.

I vividly remember the winter of 1962–1963. It was one of
the coldest on record. Rivers, lakes and even the sea froze.
I was fortunate enough to be at a school where the teachers
were genuinely interested in their pupils: not just from an
educational point of view, but from a personal development
angle as well. Our headmaster was a silver-haired gent called
Charles Mcgregor, a Scot with an artificial leg. He was one of
those post-war teachers who, having seen action in the
Second World War, was now in a leadership role with us
kids. He was a marvellous master and my thoughts of him are
filled with affection and respect. He had teachers in his
mould too, and they willingly gave up their spare time to us
on outdoor hikes and expeditions.

I mention this because it was in early 1963 that a party of
us, accompanied by dedicated teachers, went to Borrowdale
in the Lake District for a long weekend. We stayed in the
Longthwaite Youth Hostel and climbed Great Gable and
Scafell Pike in the snow and ice. It was marvellous; it was exciting; it was adventurous. This would never happen today:

adventure is frowned upon unless it meets strict safety standards, which kind of defeats the object. The highlight of our
trip was walking right across Derwent Water on the thick ice.

That is my abiding memory of the winter of 1962–1963,
but what a winter it was across the United Kingdom. The
months of December, January and February were known as
the 'big freeze'. Temperatures dropped to −22 °C. Snow
blocked roads and railway lines and communities were cut
off for days and weeks on end. Strangely, I don't remember
my school being closed for a single day. There was no mention of 'climate change'. People just got on with it and knew
that as soon as spring came along the weather would improve, which it did.

Fast forward to the summer of 1976. It was considered
the hottest summer in Europe during the 20th century. High
pressure moved in during late May and stayed there until the
first traces of rain on August 22. Three months of scorching
weather. During this spell, temperatures exceeded 32 °C at
several weather stations every day for three months; Chelten-
ham had 11 successive days of 35 °C. Roads melted, rivers and
reservoirs dried up and standpipes were introduced at various locations. Yorkshire alone had 11,500 of these pipes as
people queued for water.

The slogan 'Save water, bath with a friend' appeared
everywhere and caused much mirth. This 16-week dry spell
was the longest recorded over England and Wales since 1727.
Large tracts of countryside were cordoned off and the public
were not allowed to walk or hike because of the risk of fires,
though many did break out and destroyed trees, moorland
and property. From mid-summer on, wild fires became a national preoccupation and the news was dominated by the
spectacular accounts of 'pyrotechnics' when Surrey heaths
and the North York Moors went up in smoke.

Urban areas suffered in their own way. The London Underground was hell on earth; office work became an ordeal without air conditioning, and to cap it all that annual strawberry fest at Wimbledon resulted in 400 spectators being treated for heat exhaustion in a single day. It was so bad that stewards at Wimbledon and Henley Regatta were allowed to remove their jackets—incredible, I know, but true.

Was there any talk of climate change? Did the government scare us silly with prophecies of doom and gloom? I don't think so. Yes, it was a belting summer and one which has been as a benchmark for comparison ever since. Which is why when you look back not too long ago, the winters and summers of recent years have been mild and not worthy of note, unless of course the powers that be have a different agenda.

So to the 2022 'summer heatwave'. What a joke! It beggars belief, I know, but the government, via the NHS, produced 'a heatwave plan for 2022' and because this was nothing short of a damp squib they thought they would carry it forward to 2023.[1] And what of 2024 I hear you ask? Have no fear, there is pressure to bear for more 'heatwave hysteria' and none other than that bastion of all things nonsensical—*The Guardian*—are calling for action to save lives.[2] Like the puffs of wind that now have names they are calling for heatwaves to be assigned a named identity. Who decides this stuff? Is this what our taxes pay for? How many people are sitting at their computers and formulating this drivel? I think it might be a good

---

1 'Adverse Weather and Health Plan, and summer 2023 preparedness—protecting communities and public services from adverse weather', 8 June 2023, *NHS England* [website], <https://england.nhs.uk/publication/adverse-weather-and-health-plan-and-summer-2023-preparedness-protecting-communities-and-public-services-from-adverse-weather>, accessed 30 August 2024.

2 Damian Carrington, 'UK heatwave plan urgently needed to save lives, say MPs', 31 January 2024, *The Guardian* [website], <https://theguardian.com/environment/2024/jan/31/uk-heatwave-plan-urgently-needed-to-save-lives-say-mps>, accessed 30 August 2024.

idea if they gave up on heatwave plans and concentrated on getting the National Health Service back to a state whereby you can actually see a doctor face to face, but that seems like a pipe dream. Not to mention that if the government are concerned about deaths due to temperature fluctuations, they would be far better reversing their policy of impoverishing millions of people by their attacks on living standards. Perhaps this would then free up countless numbers during the winter months of having to choose whether to 'heat or eat'. I'm afraid far more people in the UK suffer from the cold than they ever do from the heat, but that is far to obvious for the powers that be to comprehend. And it doesn't chime with their narrative.

The winter of 1963, the summer of 1976, exceptional years no doubt, but recent years? Bog-standard by any measure, yet used as another weapon in the war to scare the population witless. 'Climate change' is a nonsense and for those of us who were around in the Sixties and Seventies, it is blatantly obvious that the whole climate agenda is being used for nefarious purposes. The government would have us go from one disaster to another, when all we are faced with is life as it unfolds, just as it has done for millennia.

Within weeks of being re-elected mayor of London, Sadiq Khan has drawn up plans to charge electric car owner £15 per day to drive in London from next year to save the planet from climate change. I understood electric cars did not cause CO_2 emissions. This proves that these charges are not about saving the planet or climate change, but raising cash. Of course, he with his 6-litre gas guzzler will be exempt. Also, China are selling thousands of solar panels to us, yet we are not told these panels are being made in oil and coal-fired factories.

—*John Willis*

XXXIII.

A modest home experiment—extra CO_2 does not give more warming

By Auralaywales

Auralaywales has a Physics degree and has worked in industry before becoming a science teacher. Now retired, he enjoys DIY in the house and is interested in anything technical, from Astronomy to Monetary Theory.

In the early 2000s the ecology movement started claiming that all the CO_2 we were making would dangerously warm the planet, because of The Greenhouse Effect. The answer was to stop burning fossil fuels and harness renewable energy. Wind turbines started appearing on hilltops, later solar panels on roofs. Many true ~~greens~~ lovers of nature suddenly saw how intrusive they looked and switched to complaining about the destruction of the landscape.

I have to admit that I found this 180 degree turn rather funny, until I found out how much my electricity bills would rise to subsidise them.

Then I learnt that whatever the rated power of the turbines, in real life the output varied with the weather; they would produce on average about a third of their theoretical output. Piling on the absurdity, this required that conventional generators—usually gas—had to be kept on standby, ready to start up when the wind didn't blow. This was doubly inefficient, first by keeping the plants staffed and ticking over just to be ready when needed, and also because the kind of generator that can be ramped up and down rapidly is far more wasteful of fuel than one designed to run at constant power.

At first, in my naivety, I thought that common sense, based on genuine science, would win through. My hopes were especially raised when a leak of emails and program code from the University of East Anglia showed the scientists were conspiring to suppress any contrary views, and that the computer models were a hopeless amateur mishmash with no real world validity.

I expected the whole edifice, built on sand, would collapse. Of course, it hasn't.

I see now that while it started with science, it became pure propaganda, fronting for money-making scams. There are now so many vested interests, making so much money, that quiet, reasoned arguments have no effect. The demonisation of CO_2 as causing global warming/ heating/ boiling is so embedded in our culture that even opponents of wind turbines, EVs, heat pumps, etc. have to say they believe we must cut CO_2, just that those particular contraptions are not the answer.

Yes, one day mankind will run out of fossil fuels, but that is still far in the future. We have plenty of time to develop long term solutions like nuclear fusion and practicable batteries for vehicles.

'Men, it has been well said, think in herds; it will be seen that they go mad in herds, while they only recover their senses slowly, one by one.'—Charles MacKay, 'Extraordinary Popular Delusions and the Madness of Crowds', 1841.

We have to educate people into what is going on, one by one. Rebuild understanding from the foundation up, brick by brick.

I would like to propose a little home experiment to show the effect of increasing CO_2 in the atmosphere.

You will need a pack of coloured cellophane, this can be bought cheaply from a craft shop or online, and four objects with bright primary colours: red, yellow green and blue. Lego, for instance, is perfect.

A quick digression into the greenhouse theory. Very hot objects like the Sun emit a lot of radiation in many wavelengths. Visible light plus some that is invisible, short wave ultra violet and longer wave infra red. Our atmosphere, including CO_2, is transparent to visible light and some infra red. Some radiation is reflected back by clouds, the rest comes down to the surface and warms the ground.

The warm ground also emits radiation, but much weaker and at much longer wavelengths than the sun. Some of this radiation escapes to space and cools the Earth. However, our atmosphere is less transparent to this radiation. CO_2, for instance, is 'coloured' in the infra red. It absorbs some wavelengths and lets others through. By absorbing the radiation the CO_2 warms up. If nothing else changes, the planet surface gets warmer. (By the way, that little 'if' has huge implications!) Actual measurements show that the Earth's average temperature is indeed rising slightly: 1 to 1.5 °C in the last 150 years.

Now to cellophanes. Take a red sheet and look through it. Everything goes ruby coloured because the red film absorbs blue and green light. Look at the Lego. The blue is noticeably darker. Green is a mixture of blue and yellow and so is somewhat darker. The yellow hardly changes and the red actually looks brighter against the red background.

This represents a little CO_2 in the atmosphere, say 100ppm.

(Fun fact: 100 ppm means 100 molecules in a million. If a big sports stadium of 100,000 people represents the composition of the atmosphere, just 10 of them will be CO_2 at 100ppm.)

Double the sheet and then double again, looking at the effect. We have gone to 200ppm, and on to the current 400 ppm CO_2—there are now 40 CO_2 people in the stadium, scattered amongst the 100,000.

The blue now looks black and the green isn't getting any darker. The red cellophane, representing CO_2 in the air has absorbed all the blue colour there is. Adding more layers of red makes very little difference to the blue. It already looks black, and cannot get any darker. We say the red colour is saturated. (The cellophane is not perfectly transparent to red and adding more layers does make the background look darker, which is why I want to concentrate on the blue.)

In the same way, CO_2 is pretty much saturated at 400ppm. It is already absorbing nearly 90% of all the energy it can.

Adding more CO_2 will cause very little further warming of the Earth.

This means there is no point in stopping the use of of fossil fuels—the extra CO_2 is not going to have any more effect on the temperature! Here a good article to start looking further: [1]. This a little more technical: [2].

---

1 Chris Morrison, 'New Scientific Evidence That CO_2 Emissions Can't Warm Atmosphere Because it is "Saturated" Published in Peer-Reviewed Journal', 24 April 2024, *The Daily Sceptic* [website], <https://dailysceptic.org/2024/04/24/new-scientific-evidence-that-co2-emissions-cant-warm-atmosphere-because-it-is-saturated-published-in-peer-reviewed-journal/?highlight=co2%20saturation>, accessed 30 August 2024.

2 Chris Morrison, 'According to the Work of Two Distinguished Atmospheric Scientists, Net Zero is Completely Pointless', 8 March 2023, *The Daily Sceptic* [website], <https://dailysceptic.org/2023/03/08/according-to-two-distinguished-atmospheric-scientists-net-zero-is-completely-pointless/?highlight=co2%20saturation>, accessed 30 August 2024.

Please try this with your 'true believer' friends. Tell them that this is real science.

You have taken a theory (that adding extra layers will saturate the colour), tested it by experiment and drawn conclusions.

Of course they won't believe you and they won't thank you for it. But you may just have sown a few seeds of doubt.

Next time, you can tell them that CO_2 is plant food.

300 or so million years ago, CO_2 was much higher than today. Plants flourished, as did the animals feeding on them. Then trees became so huge and prolific that they sucked a lot of the CO_2 out of the air and stored it as carbon compounds. When they died, the carbon was buried and fossilised to coal.

Likewise, many marine plants and animals died, were buried in the mud and became oil.

CO_2 became much more scarce, plants starved, and did not grow so well.

Today, when we burn coal and oil we are returning some of that buried CO_2 to a grateful nature.

More CO_2 means plants grow better and there is more food for people and animals.

The planet is getting greener and this can be seen in space pictures from NASA.

But keep that lesson for another day. Softly softly, one step at a time!

In 1989 the UN senior official predicted that entire nations would be wiped off the earth by rising seal levels as they would rise by three feet due to melting polar ice caps and would wipe out the Maldives and other flat islands by what was then called the greenhouse effect (later called global warming and now climate change, but it's still the same thing) and we had just 10 years to save the planet. That was 35 years ago. We were told it was caused by aerosols in perfume and hairspray, gas in our fridges, leaded fuel, fossil fuel (oil is not officially a fossil fuel). We got rid of those things and nothing changed, so now they constantly look for new things to blame, like cow burps. For the past ten years I have read news reports and rubbish from the BBC how many parts of America will be underwater within the next few years, and I am still waiting to see it. The only reason we get floods now, is due to building on flood plains and covering everywhere in concrete. I lived on the coast for many years and each time I return, the sea is no closer to covering that town than it was then, proving we are being lied to.

We are constantly being bombarded with more and more lies, many so stupid you just can't laugh at them any more.

—*James McGuinn*

XXXIV.

Carbon dioxide has zero impact on global warming

By Dr John Gideon Hartnett

Dr John Gideon Hartnett is an accomplished Australian physicist, cosmologist, and noted biblical creationist. A PhD holder from The University of Western Australia (UWA), Hartnett has served at both UWA and the University of Adelaide, contributing to over 200 scientific papers. He was a founding director of a successful startup that has commercialized his research on ultra-stable cryogenic 'clocks'. He's known internationally as a speaker on biblical creation, and has written extensively on the subject, especially from an astrophysics and cosmology perspective. Hartnett also writes popular science pieces, frequently critiquing man-made climate change and green energy.

The following article was kindly offered to us by Dr Hartnett for reprinting in The Green Cheat.

Carbon dioxide saturation in the atmosphere means zero impact on global temperatures, challenging the beliefs of the Climate Cult. I highlight contradictory scientific experimental evidence and criticize the insane global climate initiatives.

Ever heard of carbon dioxide saturation in the atmosphere? It means that by adding more carbon dioxide to the atmosphere above a certain threshold (of about 300 ppm; currently it is 400 ppm) can have no significant impact on global climate. This is because the atmosphere is already saturated and higher concentrations do not lead to any further absorption of radiation and hence heating.

Three Polish physicists studied this saturation principle and published three peer-reviewed papers (Kubicki et al., 2024[1], 2022[2] and 2020[3]). The latest paper included their experimental evidence.

Their 2024 paper, J. Kubicki, K. Kopczyński, J. Młyńczak, *Climatic consequences of the process of saturation of radiation absorption in gases* March 2024, published in *Applications in Engineering Science*, (Kubicki et al., 2024[1]) summarizes the results of two experiments, which they had proposed and described in their 2020 and 2022 papers: 'In the study (Humlum et al., 2013), the authors demonstrated that peaks of cyclic changes in air and water temperature globally precede peaks of cyclic changes in atmospheric CO_2 concentration (Fig. 12). This finding supports the hypothesis that, as a result of saturation processes, emitted CO_2 does not directly cause an increase in global temperature. Instead, it suggests that an increase in temperature likely leads to the release of carbon dioxide from the oceans.' (Kubicki et al., 2024)

---

1 Jan Kubicki, Krzysztof Kopczyński, Jarosław Młyńczak, 'Climatic consequences of the process of saturation of radiation absorption in gases', March 2024, *ScienceDirect* [website], <https://www.sciencedirect.com/science/article/pii/S2666496823000456>, accessed 30 August 2024.

2 Kubicki et al., 'Absorption characteristics of thermal radiation for carbon dioxide', 30 September 2022, *Lublin University of Technology—Publishing House* [website], <https://ph.pollub.pl/index.php/iapgos/article/view/2998/2729>, accessed 30 August 2024.

3 Kubicki et al., 'Saturation of the absorption of thermal radiation by atmospheric carbon dioxide', 2020, *pollub.pl* [website], <https://ph.pollub.pl/index.php/iapgos/article/view/826/1294>, accessed 30 August 2024.

Here is their Fig. 12:

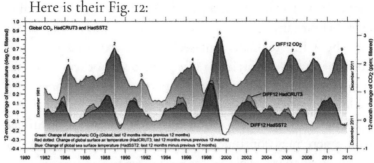

Periodic change in global atmospheric CO_2 concentration (top curve), global sea surface temperature (bottom curve), and global surface air temperature (in-between curve)

The authors are suggesting the possibility that an increase in global temperature actually causes the CO_2 concentration to increase in the atmosphere. This is the opposite of the mantra of the Climate Cult.

Furthermore they wrote: 'By comparing the saturation mass of CO_2 with the quantity of this gas in Earth's atmosphere, and analyzing the results of experiments and measurements, the need for continued and improved experimental work is suggested to ascertain whether additionally emitted carbon dioxide into the atmosphere is indeed a greenhouse gas.' (Kubicki et al., 2024)

Based on their experimental measurements they question whether CO_2 is even a greenhouse gas at all, and therefore whether it has any consequence for the fake 'climate catastrophe':

'The presented material shows that despite the fact that the majority of publications attempt to depict a catastrophic future for our planet due to the anthropogenic increase in CO_2 and its impact on Earth's climate, the shown facts raise serious doubts about this influence.' (Kubicki et al., 2024)

This is not something Climate Cult wants to hear. This new evidence is also the reason to show that so-called 'climate science' is not science but a pseudo-religious ideology leading to global tyranny.

The Climate Cult rent-seekers are extracting billions from the richer countries to line their own pockets while remonstrating they are saving the poor countries from rising sea levels.

The bottom line? Carbon dioxide concentration in the atmosphere may continue to rise but it just isn't possible for any increase to cause global temperatures to rise.

This means the effect of increasing CO_2 in the atmosphere is naturally limited and that limit was reached long ago (see their Fig.12 above). Carbon dioxide emissions have zero impact on the Earth's global temperatures.

So why all the fuss? The (Kubicki et al., 2024) paper concludes that Earth's atmosphere is already saturated with carbon dioxide. Any further addition has no effect on global temperatures.

I recently wrote in *Coal, the Creator's Providential Provision*[4] that even if you burnt all the coal on Earth, which couldn't happen, still you could not heat the planet by more than 2 degrees. Well, this new research adds a new constraint to that. It means the effect of burning all the earth's coal would have close to zero impact on global temperatures.

These results are in direct conflict with the globalist Climate Cult mantra being promoted by the United Nations-funded pseudoscience mob, that increasing CO_2 concentration leads to an increase in global temperatures.

The UN sponsored IPCC has for many years promoted The Great Global Warming Myth[5] wherein they claim 97% of

4 John Gideon Hartnett, 'Coal, the Creator's Providential Provision', 5 July 2024, *Bible Science Forum* [website], <https://biblescienceforum.com/2024/07/05/coal-the-creators-providential-provision>, accessed 30 August 2024.

5 'The Great Global Warming Myth': *Bible Science Forum* [website], <https://biblescienceforum.com/2020/02/05/the-great-global-warming-myth>, accessed 30 August 2024.

scientists agree. But it is just another lie. Cherry picking of data and other manipulation was used.

'Therefore, it is not surprising that the results in various significant works such as *Schildknecht* (2020) and *Harde* (2013), differ greatly from those presented by the IPCC, which is widely regarded as the sole reliable authority. This unequivocally suggests that the officially presented impact of anthropogenic CO_2 increase on Earth's climate is merely a hypothesis rather than a substantiated fact.' (Kubicki et al., 2024)

So ... the science is not settled.

Then why does the US taxpayer-funded NASA continue to push the global warming, even global boiling, narrative?

Everyone is in on the game. Massive magic money printing has meant much fiat currency is available for the agency and they are not going to miss out on that.

Everyone in the 'science' community looks for any twist to get climate change fiat funding irrespective of their field of study.

It is another scam but just think about it, a $15 trillion scam as demanded by Al Gore and his mates. That is the figure the globalist cult is now suggesting to fix the climate.

That is a group of crazy people who claim they want to fight global warming, published a new report calling for countries to spend up to $600 billion a year over the next two decades to boost green energy deployment and energy efficiency equipment[6] (Energy Transitions Commission's report[7]). But that is no longer on their website.

6 Geoffrey Grider, 'Al Gore's New Climate Change Group ETC Now Demanding $15 TRILLION To Save The Planet From Doom', 25 April 2017, *Now The End Begins* [website], <https://nowtheendbegins.com/al-gore-new-climate-change-group-etc-demanding-15-trillion-save-planet>, accessed 30 August 2024.

These are the key goals that the ETC believes can lead to significant mitigation of global temperatures by 2030:

> Trebling Renewables: increasing global installed renewables capacity from around 3.5 TW in 2022, up to around 11 TW by 2030.
> Energy Efficiency: doubling annual primary energy efficiency gains from ~2% p.a. up to ~4% p.a. through to 2030.
> Oil and Gas: Achieving near-zero methane and zero routine flaring by 2030.
> Carbon Capture: An aspirational goal for gigatonne-scale carbon capture capacity by 2030.
> Heavy Emitting Sectors: Launch of the Industrial Transition Accelerator (ITA) to increase low-carbon supply and accelerate demand for green products.
> Deforestation: Protect, restore and sustainably manage forest basins in key tropical countries, helping to stop global deforestation.
> Food Systems: A range of proposals to accelerate the uptake of alternative proteins, fertiliser innovations, dietary shifts, and reductions in agricultural methane emissions.

All of this is totally insane nonsense if you have not first established that CO_2 causes any increase in global temperature. Even their goals are anti-science, anti-physics. The asylum has been overtaken by the inmates.

---

7 'COP28: A High-Level Assessment of Mitigation Proposals', *Energy Transitions Commission* [website], <https://energy-transitions.org/bitesize/cop28-assessment-mitigation-proposals>, accessed 30 August 2024.

There is no existential threat in relation to the climate changing. 'Climate change' is a fraud. The climate changes naturally and various ways of temperature measuring methods show no alarming changes in temperatures globally.

Tree rings, satellites, oceans, rural, seabeds and balloons are ways of measuring temperatures. It is relevant where the temperatures are being measured as concrete in cities influences the temperature to be more hot.

The whole climate change scare hoax started in the 1970s and 1980s at the universities in the science field, which were government funded, and in politics. Climate change was just another way to suck finances out of the tax system, shame and blame people, to control us and to justify the killing of humans eventually.

—Sara Kristensen, Denmark

XXXV.

Net Zero averted temperature increase

By Dr Richard Lindzen, Dr William Happer, Dr William A. van Wijngaarden

The following article* has been kindly offered us by Dr Happer for reprinting in The Green Cheat.

> Dr Richard Lindzen is an atmospheric physicist and has published over 200 scientific papers and books. From 1983 to 2013 he was Professor of Meteorology, in the Department of Earth, Atmospheric, and Planetary Sciences at the Massachusetts Institute of Technology.
>
> Dr W. Happer is Professor Emeritus in the Department of Physics at Princeton University and a specialist in modern optics, optical and radiofrequency spectroscopy of atoms and molecules, radiation propagation in the atmosphere, and spin-polarized atoms and nuclei. He has published over 200 peer-reviewed papers.
>
> Dr W. A. van Wijngaarden joined the Department of Physics and Astronomy at York University, Canada in 1988. Most recently, his group developed an array of microtraps of ultracold atoms. He has 75 refereed publications and given over 200 conference presentations and invited seminars.

*   License: CC BY 4.0, arXiv:2406.07392v1 [physics.ao-ph] 11 Jun 2024.

Abstract

U sing feedback-free estimates of the warming by increased atmospheric carbon dioxide (CO_2) and observed rates of increase, we estimate that if the United States (U.S.) eliminated net CO_2 emissions by the year 2050, this would avert a warming of 0.0084 °C (0.015 °F), which is below our ability to accurately measure. If the entire world forced net zero CO_2 emissions by the year 2050, a warming of only 0.070 °C (0.13 °F) would be averted. If one assumes that the warming is a factor of 4 larger because of positive feedbacks, as asserted by the Intergovernmental Panel on Climate Change (IPCC), the warming averted by a net zero U.S. policy would still be very small, 0.034 °C (0.061 °F). For worldwide net zero emissions by 2050 and the 4-times larger IPCC climate sensitivity, the averted warming would be 0.28 °C (0.50 °F).

1 Introduction

In this note, we show how to simply estimate the averted temperature increase δT that would result from achieving net zero carbon dioxide emissions in the United States (U.S.) or from worldwide net-zero policies. Straightforward calculations outlined below show that eliminating U.S. CO_2 emissions by the year 2050 would avert a temperature increase of

$$\delta T = 0.0084 \,°C, \tag{1}$$

less than a hundredth of a degree centigrade.

Computer models are not needed to estimate the averted temperature increase (1). It is given to high accuracy by the simple formula

$$\delta T = S \log_2 \left(\frac{C}{C'} \right), \qquad (2)$$

where $\log_2$ denotes the base-2 logarithm function.

In (2) the symbol S denotes the equilibrium temperature increase caused by a doubling of atmospheric CO_2 concentrations. We will assume a numerical value

$$S = 0.75\,^\circ C. \qquad (3)$$

Because it is so hard to determine how much of the warming of the past two centuries has been from natural causes and how much is due to increasing concentrations of greenhouse gases, it is not possible to obtain a reliable estimate of S from observations. The value (3) is a straightforward, feedback-free estimate that comes from the basic physics of radiation transfer. For example, see p. 19 in the recent review of climate sensitivities[1]. The value (3) is almost the same as the estimate of Rasool and Schneider[2], S=0.8 C in the year 1971, before global-warming alarmism became fashionable.

In (2) the symbol C denotes the concentration of atmospheric CO_2 in the net-zero target year 2050 if the U.S. takes no measures to reduce emissions. The symbol C' is the concentration if the U.S. reduces its emissions to zero at that time. The U.S. fraction f_0 of total world emissions CO_2 in the year 2024 is very nearly[3]

$$f_0 = 0.12, \qquad (4)$$

1 R. Lindzen, 'On Climate Sensitivity', December 2019, *CO₂ Coalition* [website], <https://co2coalition.org/wp-content/uploads/2021/08/On-Climate-Sensitivity.pdf>, accessed 23 August 2024.

2 S. I. Rasool and S. H. Schneider, 'Atmospheric Carbon Dioxide and Aerosols: Effects of Large Increases on Global Climate', Science 173, 138–141; *ISTOR* [website], <https://www.jstor.org/stable/1732207>, accessed 23 August 2024.

3 United States Emissions of CO₂, *Our World in Data* [website], <https://ourworldindata.org/co2/country/united-states#what-share-of-global-co2-emissions-are-emitted-by-the-country>, accessed 23 August 2024.

12% or about 5 out of 40 billion metric tons of CO_2. Most emissions now are from China and India. Therefore the concentration decrement, δC, if the U.S. reduces emissions to zero by the year 2050,

$$\delta C = C - C',\qquad(5)$$

will be relatively small,

$$\frac{\delta C}{C} \ll 1.\qquad(6)$$

We can use (6) to approximate (2) as

$$\delta T = -S\log_2\left(1-\frac{\delta C}{C}\right)$$

$$\approx \frac{S\,\delta C}{\ln(2)C}$$

$$\approx \frac{S f_0 R \Delta t}{2\ln(2)(C_0 + R\Delta t)}.\qquad(7)$$

Before turning to the derivation of (7), which assumes the U.S. fraction of world emissions decreases steadily from $f_0=0.12$ now to zero in the year 2050, we discuss the meanings of the symbols and we give representative values of them. The natural (base-e) logarithm of 2, which appears in (7), has the numerical value

$$\ln(2)=0.6931.\qquad(8)$$

The atmospheric concentration of CO_2 now (the middle of the year 2024) is[4]

$$C_0 = 427\text{ppm}.\qquad(9)$$

The time remaining to the net zero target date of 2050 is

$$\Delta t = 25.5\,year,\qquad(10)$$

4 CO_2 Concentration in the Year 2024; *Global Monitoring Laboratory—Earth System Research Laboratories* [website], <https://gml.noaa.gov/ccgg/trends/monthly.html>, accessed 23 August 2024.

The current rate of increase of atmospheric concentrations of CO_2 is

$$R = 2.5\text{ppm } year^{-1}. \tag{11}$$

Substituting numerical values from (3), (4), (8), (9), (10) and (11) into the bottom line of (7) gives (1).

2 DETAILS

If there were no reductions of the U.S. fraction of CO_2 emissions, the atmospheric concentration at the net zero target date would be

$$\begin{aligned} C &= C_0 + \Delta C \\ &= 490.75\text{ppm.} \end{aligned} \tag{12}$$

If the emission rate continues at the constant value R for the time Δt the concentration increment would be

$$\begin{aligned} \Delta C &= R \Delta t \\ &= 63.75\text{ppm.} \end{aligned} \tag{13}$$

We used (10) and (11) to write the bottom line of (13), and we used the bottom line of (13) with (9) to write the bottom line of (12). Because the radiative forcing of CO_2 is proportional to the logarithm of the concentration, the temperature increment in the year 2050, caused by the concentration increment (13), would be

$$\begin{aligned} \Delta T &= S \log_2 \left| \frac{C}{C_0} \right| \\ &= 0.1506\,°C. \end{aligned} \tag{14}$$

The numerical values of S from (3), of C_0 from (9) and C from the bottom line of (12) were used to evaluate the bottom line of (14).

The proportionality of the temperature increment ΔT to the logarithm of the concentration ratio C/C_0 means that the warming from increased CO_2 concentrations C is 'saturated.'

That is, each increment dC of CO_2 concentration causes less warming than the previous equal increment. Greenhouse warming from CO_2 is subject to the law of diminishing returns.

If the U.S. continued to contribute the same fraction f_0 of (4) to world CO_2 emissions between now and the net zero target date, the U.S. contribution to (13) would be $f_0 R \Delta t = 7.65$ ppm. But if the U.S. fraction of emissions decreased steadily to zero in the year 2050, the concentration decrement (5) would be

$$\delta C = \int_0^{\Delta t} dt\, R f_0 \left(1 - \frac{t}{\Delta t}\right)$$

$$= \frac{1}{2} f_0 R \Delta t$$

$$= 3.83\text{ppm}. \tag{15}$$

We used the numerical values of (4) and (13) to evaluate the bottom line of (15). Compared to the increase ΔT of (14), the temperature would increase by a slightly smaller amount for a U.S. net zero scenario,

$$\Delta T' = S \log_2 \left| \frac{C - \delta C}{C_0} \right|$$

$$= 0.1421\,°C. \tag{16}$$

The averted temperature increase δT from net-zero policies is

$$\delta T' = \Delta T - \Delta T'$$

$$= 0.0085\,°C. \tag{17}$$

The bottom line of (17) came from subtracting the bottom line of (16) from the bottom line of (14).

We can use the top lines of (14) and (16) to find a convenient formula for δT

$$\delta T = \Delta T - \Delta T' = S\left[\log_2\left(\frac{C}{C_0}\right) - \log_2\left(\frac{C-\delta C}{C_0}\right)\right]$$

$$= S\log_2\left(\frac{C}{C-\delta C}\right)$$

$$= -S\log_2\left(1-\frac{\delta C}{C}\right). \tag{18}$$

Recall that the base-2 logarithm, $\log_2(x)$, of some number x is related to the base-e (natural) logarithm, $\ln(x)$, by

$$\log_2(x)=\frac{\ln(x)}{\ln(2)}. \tag{19}$$

Using the power-series expansion

$$-\ln(1-r)=r+\frac{r^2}{2}+\frac{r^3}{3}+\frac{r^4}{4}+\cdots \tag{20}$$

with the last line of (18) we find

$$\delta T = \frac{S}{\ln(2)}\left[\left(\frac{\delta C}{C}\right)+\frac{1}{2}\left(\frac{\delta C}{C}\right)^2+\frac{1}{3}\left(\frac{\delta C}{C}\right)^3+\cdots\right]$$

$$\approx \frac{S}{\ln(2)}\left(\frac{\delta C}{C}\right)$$

$$\approx \frac{Sf_0R\Delta t}{2\ln(2)(C_0+R\Delta t)}. \tag{21}$$

Because of (6), each term on the right of the first line of (21) is at least 100 times smaller than the previous one. So the first term is a good approximation to the sum. The value from the approximate formula on the second or third line of (21) only differs by about 1% from the exact value of δT, which is given by the sum of the infinite number of terms on first line. Eq. (21) completes the derivation of (7).

The Green Cheat

3 ALTERNATE ASSUMPTIONS

Using the last line of (7), we can see what happens if we use alternate assumptions about the averted temperature increase. For many years the United Nations Intergovernmental Panel on Climate Change (IPCC) asserted that the most likely value of the equilibrium climate sensitivity is four times larger than the feedback-free value (3),

$$S = 3.0\,°C. \tag{22}$$

This assumes a positive feedback that increases the warming by 400%. According to Le Chatelier's principle, most feedbacks in nature are negative. But if we use the dubious value (22) in (7) we find that the U.S. net zero scenario would avert a temperature increase of

$$\delta T = 0.034\,°C, \tag{23}$$

less than four hundredth of a degree centigrade.

As less developed countries use fossil fuels to raise their standards of living, it is reasonable to expect that the rate of growth of atmospheric CO_2 will increase above the current value, even if the U.S. and other countries implement net zero policies. Suppose the growth rate increases by 30% from the current value of (11) to

$$R = 3.25\text{ppm}\,year^{-1}. \tag{24}$$

If we use the value (24) in (7) we find that driving U.S. CO_2 emissions to zero by the year 2050 would avert a temperature increase of

$$\delta T = 0.011\,°C, \tag{25}$$

slightly more than one hundredth of a degree centigrade.

The temperature increment (25) was estimated for the physically reasonable climate sensitivity $S=0.75$ °C of (3), and the growth rate $R=3.25$ ppm year^{-1} of (24) that is 30% larger than the current growth rate $R=2.5$ ppm year^{-1} of (11). If we use IPCC's 4-times larger, but dubious climate sensitivity

S=3.0 °C of (22), along with the larger growth rate R=3.25 ppm year^{-1} of (24), we find an averted temperature increase of

$$\delta T = 0.042\,°C, \qquad (26)$$

slightly more than four hundredth of a degree centigrade.

4 Worldwide Net Zero

We can calculate the averted temperature increase, δT, if the entire world adopted net zero policies and reduced their CO_2 emissions to zero by the year 2050. Then the formula for the averted temperature increase would be given by (7) with the fraction f_0=1,

$$\delta T = \frac{SR\,\Delta t}{2\ln(2)(C_0 + R\,\Delta t)}$$

$$= 0.070\,°C. \qquad (27)$$

The numerical value of the second line comes from evaluating the expression with the most likely numerical values of (3), (8), (9), (10) and (11).

Using the four-times larger sensitivity S=3 °C of (22) instead of the more physically reasonable value, S=0.75 °C of (3) to evaluate (27) we find an averted temperature increase of

$$\delta T = 0.28\ °C. \qquad (28)$$

5 The MAGICC model

In a prepared statement before the U.S. Senate Budget Committee, B. Zycher[5] showed that the MAGICC model (Model for the Assessment of Greenhouse Gas Induced Climate Change)[6], projects that if the U.S. reduced emissions to zero

5 B. Zycher, formal statement before U.S. Senate Committee on the Budget, 29 March 2023; [file], <https://budget.senate.gov/imo/media/doc/Dr.%20Benjamin%20Zycher%20-%20Testimony%20-%20Senate%20Budget%20Committee.pdf>, accessed 23 August 2024.

6 The MAGICC Model, The climate system in a nutshell, *MAGICC* [website], <https://magicc.org>, accessed 23 August 2024.

in the year 2050, the averted temperature increase in the year 2100 would be

$$\delta T = 0.173 \,°C. \tag{29}$$

The time to net zero for this scenario would be

$$\Delta t = 75.5 \, year, \tag{30}$$

instead of $\Delta t = 25.5$ year as in (10). Zycher used an even larger climate sensitivity

$$S = 4.5 \,°C, \tag{31}$$

than the value, $S = 3.0 \,°C$ of (22). From inspection of (15) we see if net US emissions were reduced to zero in a shorter shorter time

$$\Delta t_{us} = 25.5 \, year, \tag{32}$$

than the time $\Delta t = 75.5$ years until the year 2100, the averted concentration increment in the year 2100 would be

$$
\begin{aligned}
\delta C &= R f_0 \Delta t - \int_0^{\Delta t_{us}} dt \, R f_0 \left(1 - \frac{t}{\Delta t_{us}}\right) \\
&= R f_0 \left(\Delta t - \frac{1}{2} \Delta t_{us}\right) \\
&= 18.8 \text{ppm}. \tag{33}
\end{aligned}
$$

a factor of about 5 larger than (15) because of the long, 50-year interval from 2050 to 2100 of net zero U.S. emissions. The numerical value on the bottom line of (33) was evaluated with (4), (10), (11) and (32).

Substituting (33) into (21) we find

$$
\begin{aligned}
\delta T &= \frac{S}{\ln(2)} \left(\frac{\delta C}{C}\right) \\
&= \frac{S f_0 R (2\Delta t - \Delta t_{us})}{2\ln(2)(C_0 + R\Delta t)} \\
&= 0.20 \,°C. \tag{34}
\end{aligned}
$$

The numerical value on the bottom line of (34) is reasonably close to the MAGICC estimate (29). It was evaluated with the parameter values from (4), (9), (11) and (30)–(32).

6 CONCLUSION

As shown by (1), (23), (25) and (26), there appears to be no credible scenario where driving U.S. emissions of CO_2 to zero by the year 2050 would avert a temperature increase of more than a few hundredths of a degree centigrade. The immense costs and sacrifices involved would lead to a reduction in warming approximately equal to the measurement uncertainty. It would be hard to find a better example of a policy of all pain and no gain.

Hope Amidst a Tsunami of Evil—Exposing the Great Lies. Veronica Finch.

2022, paperback, 250 pages, £15.

Evidence, facts, essays, testimonies, letters and reflections.

The White Rose—Defending Freedom. Veronica Finch.

2021, paperback, 102 pages, £10.

All about the White Rose UK, including English translations of flyers from the German underground resistance.

Freedom!—An Anthology of Poems, Short Stories and Essays. Various authors.

2021, paperback, 147 pages, £10.

Composed by 36 authors during Covid restrictions and lockdowns.

The Big Bad Wolf and the Syringe —A Fairy Tale. Connie Lamb.

2022, 35 pages, online: £7, offline: £5 incl. p&p.

Can the hens of Henton trust Wolf's 'miracle cure'?

Stop Them!—Together We Will End World Control, and This Is How!

Propositions from 81 authors. 2023, 50 pages, free download.

How To Avoid Digital Slavery is filled with exciting fiction and valuable advice from forty authors. This 300 pages long publication will help you take back control and live a healthy life in freedom. 2023, hardcover, £20.

Prayer Is Power, Download this booklet for free or purchase the printed booklet, online: £3, offline: 2 for £5.

Freedom Magazine—Single and back issues; subscribe for £6 (UK), £9 (international) per month incl. p&p. Contact us for bulk orders: magazine.thewhiterose.uk.

Order online: thewhiterose.uk. We also accept cash payments (banknotes only—no cheques please): TWR, PO Box 1482, PRESTON, PR2 0ER. Free p&p in the UK. Overseas customers: please add voluntary payment for p&p in banknotes (prayer booklet, Big Bad Wolf and magazines excluded).